Edwin Lester Linden Arnold

**Bird Life in England**

Edwin Lester Linden Arnold

**Bird Life in England**

ISBN/EAN: 9783337095444

Printed in Europe, USA, Canada, Australia, Japan

Cover: Foto ©berggeist007 / pixelio.de

More available books at **www.hansebooks.com**

# BIRD LIFE IN ENGLAND

BY

EDWIN LESTER ARNOLD

AUTHOR OF "ON THE INDIAN HILLS," "A SUMMER HOLIDAY IN SCANDINAVIA,"
"COFFEE PLANTING IN INDIA," ETC.

London
CHATTO AND WINDUS, PICCADILLY
1887

TO

MY FATHER,

MY COUNSELLOR AND COMPANION

IN HOURS BOTH OF WORK AND PLAY,

I GRATEFULLY DEDICATE

ALL THAT IS LEAST UNWORTHY IN

THESE PAGES.

# INTRODUCTION.

In these pages I have attempted to give a somewhat unconventional view of bird life in England and her sister kingdoms to north and west. The naturalist may smile at a monograph so incomplete as this, but in Yarrel, Morris, Gould, Grey, and a score of others, he will find exhaustive authors who have gone with infinite care and pains through the whole list of British birds and epitomized each. No attempt of this sort has been made here. There are, to begin with, more than three hundred birds nesting with greater or lesser frequency in our islands, and even at such modest allotment as two pages to each, we should have at once a bulky volume of six hundred pages. But the greater part of these wild fowl of meadow and marsh are the curator's birds alone, and lovers of country side and the life of copse and dingle can only hope to meet with but a very much reduced selection from the formidable list which professors and students have put together. It is of these birds, the more familiar ones, I write.

Nor, though loving the gun and a long day in the open over heather or rushes, illogical as it may seem to the uninitiated, as much as I love the birds themselves, do I pretend this volume to be any rival to Colonel Hawker's immortal "Hints to Young Sportsmen," Daniel's comprehensive "Rural Sports," Folkard's "Wildfowler," Sir Ralph Payne-Gallwey's chatty reminiscences of Irish sport, or any of some hundred volumes which authors reputable with gun and pen have given to the world. For such treatises I have neither time nor inclination. Personally, it is my opinion that not very much real learning for our guidance in the field is to be picked up in the hard and fast instruction of type.

A single season's successes and disappointments in the open, with a trusty weapon and a faithful dog at heel, will teach more about sport and wild birds than a year's rummaging in the home library and patient perusal of authors who never know an empty day or draw a cover blank. What judicious books *can* do is to foster a love for those outdoor exercises which in turn foster that spirit of resolution and patience, and strength of wind and limb, which is one of the happiest distinctions of Englishmen.

Who is there that has not read St. John's wonderful descriptions of wild fowl clamouring at night in the shallows of Scotch salt-water lochs without ardently desiring himself to hear those motley multitudes feeding in with the tide, and to learn to distinguish their infinitely varied voices—a language in itself! Or, who is there who has not burned to stalk a "muckle hart" in highland fastnesses after a dip into the seductive pages of Scrope or William Black?

At best, however, our sporting authors can only teach us

technicalities and nourish a love of out-o'-door life, all else we must learn for ourselves. Nature is the only professor to her own great lore of land and water; her lectures must be attended personally, and in her own open-air classrooms.

Such sketches of English shooting as I have included are suggestive rather than exhaustive outlines of the phases of sport they touch upon; if there is much in them that is unorthodox, it is just in that I take special pride. For some of the worst results of British game preserving every naturalist must have an utter abhorrence. I would as soon sit on a woodside gate for an hour of a summer evening waiting with a "rook" rifle ready at hand for a rabbit feeding out with the twilight from the hazels, as "grass" a hamper of pheasants outside any Midland autumn coppice.

These notes should be acceptable also to the practical agriculturist of to-day, who is surely now wiser than his ancestors were, and willing to lighten and enliven his somewhat monotonous work by observation of, and a kindly feeling for, the birds with which his vocation brings him in constant contact. It is thus I have amused myself, indeed, for lonely months and years when all other recreation was hard to come at; and these observations on bird migration, habits, and whims, have made pleasant in all weathers the monotony of Suffolk stubbles, the wild grass moorlands the Hampshire herdsmen love, and the bleak highland straths, purple in summer and white and cheerless in winter!

Country gentlemen within the last twenty years have

greatly increased their interest in and knowledge of ornithology. There is no reason why this useful, even important, knowledge should not descend to landholders in every degree, and then a good time for the bird will be at hand.

The notes I have collected on the arts of trapping and snaring are very curious and far spread. For permission to reproduce many of them I must thank our leading sporting papers, *The Field, The Sporting and Dramatic News, Land and Water, Bell's Life*, and others, to whom also I owe many and sincere thanks for the indulgence with which they have treated my frequent trespassings on their space.

A chapter on grouse moors and deer forests has been kindly supplied by J. W. Brodie Innes, Esq., Barrister-at-law.

*June*, 1887.

# CONTENTS.

# BIRD LIFE IN ENGLAND.

## CHAPTER I.

## *HAWKS AND OWLS.*

### THEIR USE AND MISUSE.

THOSE tyrants of the middle air, the falcons and their kind, may boast of as much antiquity or as remote a genealogy as any birds we know of. Over the portals of Assyrian temples we recognize the familiar hooked beak and commanding wings, and upon mummy cases of Egyptian princesses who breathed when the world was six thousand years younger than it is now the hawk comes out again aggressive and predominant.

In their relation to mankind they have played many parts. There is first of all this typification and poetic emblemism, springing from their keen vision, matchless flight, and fierce courage. This placed the eagle at Jove's footstool, and suggested him as a fitting bird to ride upon the standards of ancient Rome round the world. Even to-day, our great republics have this imperious and regal bird as their token. Strange anomaly which perches this autocrat and tyrant of his kind upon the banners of "universal equality!" As for the poets, it is difficult to say what would have happened had they been unable to draw metaphor and simile from these many-virtued wild fowl; and half our best families, half the best families in Europe, have a hawk of one kind or another for their crest.

This suggests the second count upon which the kind have proved useful to men, that, namely, wherein they have been associated with him in the pursuit of game—another link of much antiquity.

The kings of Babylon and Nineveh no doubt knew something of hawking, while the science is also of immeasurable antiquity in China and amongst the wild tribes of Central Asia. With the Normans and their conquest the art was brought to a fine, chivalrous perfection in this country; franklin and baron maintaining their own falconers and mews, while troubadours filled ballad and ditty with allusions to "gay goshawks," "gentle faucons," or their kindred. Then, we take it, it was a good time for all the race, as much in fashion as their quarry, the pheasants, partridges, and grouse are to-day.

Next to heraldry, falconry ranked in esteem under Normans and those who followed them. Only certain kinds of birds might be kept by certain subjects. For kings there was the ger-falcon; for princes, the falcon-gentle, or tercel-gentle; dukes had the rock-falcon, and earls the peregrine; a baron might use the bastard-falcon, and a knight the sacre and sacret; esquires, harrier or lanneret; a lady, the merlin; "young men" had the hobby, and yeomen the goshawk; the tercel was for a poor man, and sparrow-hawk for a priest; "the musket for a holy-water clerk, and kestrel for a knave."

This baronial hawking must have been pleasant enough sport then, when the Thames ran through eternal groves of oak and hazel, and the Severn shone in sunlight through far-stretching birch coppices; while the Midland and Western counties, with their great fen lands and open downs, supplied ample preserves for the wild-fowl to fly the falcons at. But to-day hawking is in abeyance—the pastime of a wise few who are, moreover, fortunate enough to live where it can be practised, on the borders of wide Hampshire downs, or vast fenceless pastures of such counties as Northamptonshire and the wolds beyond the Humber. Abroad, in India or elsewhere, it might well be followed more keenly than it is.

The economical value of vultures and kites as scavengers or removers of "matter in the wrong place," is a consideration of very little weight to the Englishman of to-day. He leaves it to his Commissioners of Sewers, and they leave it to the noble river that stinks through his capital; so the vultures are not wanted! Abroad they are invaluable in this respect, and lesser hawks keep lizards, snakes, and frogs within due limits of reproductiveness.

Almost the only interest attaching to hawks at the present time is with regard to game and game preserving. What a debt of ill-will should the falcons owe to this comparatively new hobby that has been their ruin! We can imagine a kite, in days when kites were common, wheeling in easy circles over the open courtyard of some British-Roman villa, located on the pleasant south coast or Isle of Wight, and watching with prophetic wonder the prætorian's black-eyed children feeding and teasing a pair of pheasants, two of the few, or even the very first in the island. How well justified might have seemed the kite's harsh laugh as he took another turn in the blue to glance again at the fresh importations and calculate on the chances of a successful swoop through the open corridors! Who could have fancied those brilliant but defenceless birds with the ostentatious length of tail would sweep the enemy overhead from Saxon skies, making a kite a few hundred years subsequently a rarer sight than they were that day themselves? Yet so it has turned out, to the chagrin of naturalists, and almost the whole hierarchy of the air has gone with the gledes. I would like to suggest in these pages (and all our most recent information tends to justify us) that in many cases where hawks are ruthlessly persecuted their destruction is absolutely unnecessary, the despotic barbarity of ignorant gamekeepers and the like; whilst in those cases where a taste for game unquestionably characterizes birds of prey our advice would be to strictly limit but not to eradicate the species.

After the kites our two remaining birds of prey, the lords of their kind, have already reached a Nirvana of well-deserved protection. The golden eagle is already sacred in the Highlands, while, as far as the osprey is concerned, most lairds would rather lose their favourite piper than the gallant fish-hawk honouring the lonely mountain tarns or beetling sea cliffs by making them its home.

Next to them, in the aristocracy of bulk, is the peregrine falcon. In approaching him I feel all the doubt of the pleader who has an admirable and indomitable brigand (a pirate, perhaps, would better fit the character of this ocean-loving hawk) for a client. Perhaps the only safe course to take is to throw ourselves on the mercy of the court, to plead the picturesqueness, the gallantry, and the patriotism of the peregrine! That he likes game when game is handy I cannot honestly deny. Fixing their home on the higher peaks of some wild and lonely mountain range in the North of England or Scotland, a pair will make their eyrie where the gnarled trunk and contorted branches of a birch or spruce jut out from the cliffs, inaccessible alike from above or below, and there they cater for their noisy young with a reckless disdain of tenant's rights and the lord of the manor, which we acknowledge is shocking. There, perched on a vantage-point of rock or storm-broken timber, the peregrine will sit in the sunlight for an hour at a time watching the heather or the glades in the hazel coppices far below for grouse sunning themselves or young rabbits coming out to play. It is surely worth the price of a moor-cock or two, and an occasional leveret to watch this peerless falcon's habits, to see his irresistible and unerring swoop that rarely fails to lay a victim low, and the graceful rise which follows and saves him from such certain destruction as his drop seemed to court.

Or if he is away from his own nesting-place, and owns no watch-tower on the crags, the peregrine will scour the heather and search the higher glens with an eagerness it is curious to watch. But his distinctive and peculiar process is the calm watch from an elevated spot until a victim is espied, and then a single impetuous rush. The gulls who go up to moorlands to breed are occasionally harried by him, wild ducks and waterfowl suffering too. So deadly is the onslaught of this hawk, whose flight has been calculated to reach the figure of one hundred and twenty miles an hour, that a tiercel has been known to strike the head clean off a mallard at a single "souse," and has driven a partridge so fiercely to the ground by the shock of the encounter, that the bird has rebounded the height of a man.

I have called this wonderfully graceful bird patriotic, and he stays with us all the year round, but in the spring our indigenous peregrines are greatly augmented by arrivals from abroad. Old Belon, two hundred years ago, gives a curious account of the incredible armies of hawks and kites which he saw in the spring crossing the Thracian Bosphorus from Asia to Europe, a procession swelled by whole troops of eagles and vultures; and from Syria and Africa come many of the birds that are with us all the summer, to the delight of naturalists.

Upon the sea coast, where the peregrine is generally found now, he does no manner of harm, contenting himself with the ample booty sea-ledges or puffin-haunted caves afford.

When a pair have once selected such a fastness overlooking blue water, they and their descendants will occupy the same eyrie year after year. When falconry was at the height of its popularity, these breeding-places were all known and jealously guarded. The site of a nest was placed under especial care of the occupier of land adjoining, and they were responsible by terms of tenure for the noble birds and their offspring. The loss of a right hand was the penalty for molesting breeding birds.

Such hereditary nests there have been at Goathland in Yorkshire, and Killing Nab Scar, while "Falcon Scar," "Hawk Scar," "Eagle Cliff" and such local names no doubt mark other localities.

The peregrine likes to strike his prey in fair fight. The following curious story lays stress upon this peculiarity. "Whilst out shooting with two companions one day in January, in the county of Cork, we saw," writes a sportsman, "in the midst of a sixty-acre pasture a curious conflict going on between a peregrine and what we at first supposed to be a smaller hawk, but which subsequently proved to be a woodcock, a considerable number of which birds had been driven from the coverts by the late heavy rains into the surrounding turnip fields. As the pair seemed to be coming towards us we concealed ourselves, in hopes of getting a shot, behind a bank and hedge whence we could see the whole performance. The woodcock flew close to the ground until the hawk struck at him, when, avoiding the stoop by a sharp turn, he invariably pitched on the ground until the hawk had passed; then rising pursued his flight in the direction of a thick covet about five hundred yards distant, until again obliged to pitch in order to avoid a second attack. This continued until both came within fifty yards of the bank behind which we were concealed. Here the hawk "struck" several times at the woodcock whilst on the ground, wheeling round after each "stoop" to try again. Upon this the woodcock never attempted to fly, and we could plainly see his tactics. He was evidently tired, and, sitting well back on his tail, he presented his long bill against every onslaught of his enemy, and afforded us the strange spectacle of a woodcock—by nature a very timid bird—successfully defending himself from the attacks of a large hawk. The hawk, either fearing impalement, or, as is the habit of his species, not liking to strike his prey when on the ground, avoided coming to close quarters, and always turned aside when about a foot from the woodcock. After one of these "swoops" he wheeled a little further off than before, and of this the woodcock took immediate advantage; for, twisting over the bank and hedge, he got a good start and went at his best pace towards the covert, now about three hundred yards distant. He passed close to us, but in our anxiety to get the hawk we refrained from shooting him. The peregrine at once mounted high, but out of shot, and pursued his quarry, now half-way to his haven and going at a much greater rate than I ever thought a woodcock capable of. The hawk gained visibly, but finally the woodcock darted into the thick covert a few yards ahead of his pursuer, who did not follow him. I now thought the cock would fall an easy victim to the gun, but not so; for, getting up very wild, he made his escape after all without being fired at." Truly a hard-fought day for the woodcock!

Closely resembling the larger falcon is the common buzzard; but there is one certain way to distinguish them. While the peregrine's neck is marked by a boldly contrasted black and white collar, that of the buzzard is shaded off between the two predominant body colours. Both birds are dark above and light below. In character, however, they are very different. This latter hawk, the "puttock" of Essex hinds, is, to tell the truth, somewhat a sluggard in habits. It is even suggested by Johnson in his Dictionary "buzzard" carries some indignity with it, and Milton uses the word as equivalent to stupidity. Morris in his "British Birds" has reason on his side, however, in refusing to accept such views. Bewick, one of its earliest detractors, declares the puttock is no match for a sparrowhawk, but Bewick was unquestionably misled in this matter.

More a bird of the high woods than the cliff face or sea shore, no doubt it often takes to a rocky ledge or cleft in a "scar," but for the most part the nest is found,—or rather was once found,—in high beech trees and the like, on knolls and hillocks, in such broad and ample forests as those of Hampshire or Yorkshire. From these and similar districts there is no possible reason why the beautiful plumage and startling cry of this hawk should be any longer banished.

I am inclined to think that the species of prey most naturally sought after is the rabbit. It feeds, however, for necessity has no law, upon a great variety of other food. It destroys numberless moles of which it seems particularly fond, as well as field-mice, leverets, rats, snakes, frogs, toads, the young of game and many birds, worms, insects and newts. "The way in which the buzzard procures moles is, it is said, by watching patiently by their haunts until the moving of the earth caused by the subterraneous burrowings, points out to him their exact locality, and the knowledge thus acquired he immediately takes advantage of to their destruction. His feet, legs, and bill being often found covered with earth or mud is in this manner accounted for."

These two hawks have often roused the ire of highland and lowland keepers respectively, but we have mentioned them somewhat at length because they are amongst the commonest (if the expression may be used) of our rarer hawks. Others of their family are hardly ever seen. The red-footed falcon is a very rare visitor indeed; the goshawk keeps himself wisely to the fastnesses of the Orkneys, where he might well be allowed to take his toll of mountain hares and rabbits.

The honey buzzard is almost as rare since the villainous greed of collectors ransacked the New Forest and other breeding grounds for eggs or young "in the down." The marsh harrier is far less common than it was, though its diet affords no justification for its destruction. "It feeds itself and nestlings," says Atkinson, "with young water birds—as its name suggests. Young water birds may be found in its haunts, or young rabbits or birds, a few mice or rats doubtless being not altogether unworthy of notice to such hungry customers as four young harpies."

The hen harrier, the blue hawk, is a shade worse in the game preserver's eyes. Not very long ago I was resting under a blackthorn bush on a lonely plateau on the Cotswold Hills when a hare came cantering along the top of a stone wall, and in close attendance was an old male harrier looking brilliantly blue in the sunlight, who swept backwards and forwards, occasionally stooping at the hare, a full grown one, in a way that made him wince with fear, but never, so far as I could see, actually striking him. A worse place for repelling attacks of a hawk than a stone "dyke" we should hardly think there could be, yet the hare ambled by within a few yards, "dodging every time the falcon's wings touched him, until at last they were lost in a hollow, and though I followed at my best pace I never saw how the fray ended."

There can be little doubt but that although game is not its chief food the hen harrier will tackle anything not too manifestly beyond its powers.

Sparrowhawks and kestrels hold their own in spite of all keepers can do. This is partly, perhaps, owing to the fact that they are not regarded as quite so harmful to game as some of the larger species, and also because they breed as often as not in out of the way places,—ivy-covered cliffs, ruined towers, and the like, where they do not come conspicuously under game-rearers' attention.

Unquestionably the commonest of all our British hawks, and with little doubt the most useful, the kestrel, a hawk of no repute in the old days, is perhaps more numerous in our shires than all his kindred put together. Round by the South Foreland and the white Dover cliffs we find them in abundance, hunting mice and small birds just above high water mark. There are plenty, too, all down the valley of the Thames, and especially wherever there are scarps overhung with hazel or disused quarries. The bird does practically no harm to game. It is almost wholly a fur-hunting hawk, and the farmer who suffers them to be destroyed deserves to be overrun with field mice, and to kill something like a thousand in a day's wheat thrashing, as I have known to happen where kestrels have been abolished. Small birds of all sorts, beetles, and no doubt other kindred foods, such as worms, slugs, and even frogs perhaps, form this useful red hawk's food. I cordially commend him to an enlightened forbearance!

Nor can the sparrowhawk—that nemesis of the finches—be considered very destructive to partridge or grouse. He is that dark-coloured bird we see beating the marsh lands, or mobbed by crows and sparrows as he flies guiltily from wood to wood. Twice a-year when hedges are thick with migrants he debouches himself and feeds licentiously on whatever he will, and for the rest of the time subsists variously on the changing small birds of the seasons.

He is not popular in the chicken-yard. "A neighbouring gentleman," writes Gilbert White, "one summer had lost many of his chickens by a sparrowhawk, that came gliding down by a faggot pile and the end of his house to the place where the coops stood. The owner, inwardly vexed to see his brood thus diminishing, hung a setting-net adroitly between the pile and the house, into which the caitiff dashed and was entangled. Resentment suggested the law of retaliation; he therefore clipped the hawk's wings, cut off his talons, and, fixing a cork on his bill, threw him down amongst the brood hens. Imagination cannot paint the scene that ensued. The expressions that fear, rage, and revenge supplied were new, or, at least, such as had been unnoticed before. The exasperated matrons upbraided, they execrated, they insulted, they triumphed. In a word, they never desisted from buffeting their adversary till they had torn him in a hundred pieces." Maternal feelings, I have observed, are always extravagant.

As for those admirable birds, the hawks of the night time, their continual persecution is wanton and reckless. One correspondent thinks that since "the regular destruction of owls by gamekeepers and others, it is a notorious fact that field mice have increased to an enormous extent, so much so as in many instances to do incalculable injury to the ground crops. As an instance of the good the barn owl does in the destruction of these depredators, I will mention that on an evening during the summer I was enjoying a pipe with a farmer friend of mine, sitting near an old barn, wherein a pair of owls were then rearing their young. My attention was especially attracted to them by observing their frequent arrival, with a mouse on each occasion grasped firmly in the claw. The supper of the young family that evening consisted, to my knowledge, of seventeen mice. Can any stronger plea for the protection of this grand old English bird be urged?" And another observes of a kindred bird, the brown owl: "As an instance of the rat-destroying propensities of the brown owl, a keeper, in the employ of a large landed proprietor in an English shire, found at the entrance of a rabbit's burrow a dead rat for thirteen consecutive mornings. This burrow last summer was occupied by a pair of brown owls for breeding purposes, and they then had young ones. We can scarcely suppose that these thirteen rats were the only ones destroyed during the thirteen days; but if they were, surely an occasional leveret or other game is but a small compensation for the benefits conferred by the destruction of a rat a day. Brown owls will now be protected on this estate by the gamekeepers. May this good example be followed by others!"

Groom Napier, a first-class authority, tells us of the long-eared owl, "the food of this species in April I have ascertained to be beetles, bats, and mice.

"Barn owl. Two owls contained dormice, water rats, and bats. Three contained field mice and beetles.

"Tawny owl. Two of these birds contained bats, but in one other case a young rabbit and two mice were taken from a stomach.

"The white owl is rather destructive to the young rabbits of Abbotsleigh Down, Hunts. But we may place all the owls, on the whole, as A 1 among the farmer's friends."

His subsequent and careful investigations have more than carried out these first examinations.

The whole of our ten species of owls, most of them rare and all scarcer than they should be, deserve protection. "No other bird exceeds them in service to man, silent unobtrusive service, and we have very few birds in Britain to compare with them in beauty of plumage."

The "gamekeeper's tree," or the old-fashioned barn-door, are always shocking sights to an ornithologist who appreciates the labours of the feathered kind, and recognizes their multitudinous usefulnesses, but they never touch his feelings so deeply with their array of nailed up victims as when he notices the owls there, and knows how unjustly Ascalaphos and Nyctimene have died. I look forward to a better time for both the hawks of the day and those of the night.

The sportsman avenges on birds of prey of all kind his real or fancied injuries by the severe judgment of the gun, but where falcons are needed for training to the delightful sport, not yet quite extinct in many parts of the globe, resource must needs be had to other and ingenious methods. A bait of some sort, either a living or a dead bird, is always essential. The hawk, unlike many of his weaker brethren, is not to be allured by his vanity, credulity, amativeness, or simple gullibility; it is hunger alone that will bring him from the clouds to the netsman's toils.

The mode of capturing falcons amongst the Arabs of Syria, for instance, is as follows. Supposing the Arab to have noted some particular place in which hawks abound, such as ruins or rocky places, he provides himself with a pigeon or partridge, or any bird that they may be fond of. Fastened round its body is a very fine net, and when the sportsman has placed his decoy in some convenient spot, it is not long before its struggles attract the attention of some wandering bird of prey which swoops down upon it and is entangled in the net. The captor, who has been hiding near, then rushes out, and seizing the victim places a hood on its head, after which he carries it about with perfect safety on his shoulder. As long as deprived of sight it will make no attempt to escape; and when after some months of careful and skilful training it is flown again at game, it exhibits the greatest fondness for and faithfulness to its captor.

Hawks are caught in an ingenious but cruel manner in the Deccan. A stick, about a foot in length, is thickly daubed with bird-lime, and some small bird, generally a dove, is tied to its centre. When the hawk is seen the unhappy captive has its eyes sewed up to make it soar, and is released. The enemy pounces upon it; its wings strike the limed twig and it falls to the ground. Hawking in India is not practised to anything like the extent it might be. A cast of falcons should be in the compound of every Englishman's bungalow, and there is ample scope for the spread of a delightful pastime in many of the Indian stations where other sport is distant or hard to obtain. Natives take kindly to the amusement, and become good falconers. This is the manner in which one native "mew's man," purveyed his hawk's daily food. "Having distilled some extraordinarily sticky brown-coloured bird-lime, called 'goolur,' from juice of the burr, or great Indian fig-tree, he would endue with this adhesive compound the end of a long thin stick, exactly resembling a full-sized fishing-rod. With this weapon over his shoulder Mahomed would go forth till he met with some sparrows chattering on the eaves of a low-tiled roof. At once he would hold the rod so as to be fore-shortened towards them, and then, having got within range, he made a sudden lunge, when one or more unfortunates would infallibly be seen adhering to the end of the stick. These were removed without being killed, and their heads inserted between his fingers, with their bodies outward, till his hands looked as though he had large boxing gloves on. I learnt how to do this myself well enough to catch birds out of the hedges, but I never acquired sufficient accuracy to work amongst the roofs, where an error of half an inch would be destruction to the wand—a valuable weapon, and one difficult to replace." Our own kestrel at home is such a terrible enemy to the professional bird-catcher, pouncing down and carrying off his trained decoy birds, that the following trap has to be frequently used against him. A white napkin to attract the hawk while in the air, is spread upon the ground and fastened down at the corners with little sticks. In the centre of this is a small peg to which a live sparrow is secured with a few inches of string. Slender twigs are then placed all round the napkin, so as to prevent the hawk from attacking the decoy from any position but above. Two long and slender limed willow twigs are then lightly fixed in the ground, one at each end of the cloth, so as to form an arch over the sparrow. When the kestrel strikes down at the sparrow his wings touch and stick to these limed twigs, and as they at once fall from their positions, he rolls helplessly over and over.

Sparrowhawks are also taken very often in this way, but more commonly among lesser varieties in the famous but seldom described "square net," which is thus mentioned by Sir John Sebright: "A net, eight feet in depth, and of sufficient length to enclose a square of nine feet, is suspended by means of upright stakes, into which transverse notches are made, and on which notches the meshes of the net are loosely placed, so that as soon as a hawk strikes against it the net readily disengages itself and falls. The square enclosure is open above, and within it a living bird, usually a pigeon, is fastened as a bait. The colour of the net should assimilate as much as may be with surrounding objects, and the material should be a fine silk. The merlin, the hobby, and the sparrowhawk, may be taken in this way; but the larger varieties, viz. gere-falcon, peregrine, and goshawk, are seldom to be thus trapped, and must be captured either by the bow-net, or the hand-net." The yearly migration of hawks in Austria and elsewhere gives much opportunity for the use of such snares, and quite a trade is carried on in live falcons, or in their heads, for which antiquated municipal laws offer a premium to the conscienceless pothunter!

Sometimes these passage hawks are taken by huge hand-nets, similar in principle to the landing-nets used in fishing, but very much larger. With these the hawk is caught by the falconer, who is concealed near a pigeon tied by a string to his hand, and suffered occasionally to fly a short distance. The bird attracts the hawk, who makes a swoop, and is dexterously caught by the falconer while its attention is thus fully engaged. But one of the most successful nets in use, the bow net, has only been mentioned in two or three works, though there has been much curiosity on the subject. The method and working is so clearly given in one of Beeton's excellent little handbooks, I am tempted to reproduce it here.

"*Lanius excubitor* is the bloodthirsty shrike's classic appellation. *Excubitor*, or sentinel, applies to the bird's vigilance in watching that no other bird, savage as himself, approaches its nest. Falconers take advantage of this peculiarity of the shrike to make him useful in the practice of snaring hawks. Towards the end of the year, in October and November, the hawks are on their passage to the southern and warmer climes of Europe; and at this season the falconer can secure the most birds. He builds a low turf hut in the open country, with a small opening on one side; at about a hundred yards distance from this hut, a pigeon (usually a light-coloured one, to attract the hawk while soaring high in the air) is placed in a hole in the ground, which is covered with turf, and a string is attached to it, reaching to the hut. Another pigeon is placed in a like position on the opposite side, at the same distance from the hut. At a dozen yards from each pigeon a small bow-net is fastened to the ground, which is so arranged that the falconer can pull it over, by a small piece of iron attached to the net, and leading to the hut. The string by which the pigeon is held passes through a hole in a piece of wood driven into the ground, in the centre of a bow-net. The falconer has also a decoy-pigeon, in a string at a little distance from the hut, and half-a-dozen tame pigeons are placed on the outside of the hut, which, on the sight of a hawk, immediately take shelter within. The next, and most important adjunct in the business, is the butcher-bird. He is placed on a hillock of turf at a short distance from the hut, and is fastened by a leather thong. The falconer, however, does not sacrifice the life of his servant, but humanely makes a little hole in the turf, into which the bird can escape when it chooses. Having thus everything prepared, the falconer has nothing to do but to sit in the hut, and watch the motions of the grey shrike. Habit has sharpened the sight of this little bird, and he descries his natural enemy long before the falconer would be able to see it. At first, if a hawk is approaching, the shrike exhibits a certain uneasiness, a drawing-in of the feathers, and a fixed gaze in one direction, the meaning of which the falconer knows well. Even when the hawk is at the distance of three or four hundred yards, the butcher-bird will scream with fear, and retreat into the hole in the turf. The falconer then prepares his decoy, and draws out the pigeons where the bow-nets are placed, which, by fluttering round, soon attract the hawk, who swoops at them, and is caught in the snare. Not only does the butcher-bird give its master warning of the approach of the hawk, but lets him know the species by the greater or lesser degree of alarm which it exhibits."

That magnificent vulture of South America, the great condor of the Andes, is not exactly the kind of game that would appear to lend itself most readily to the trapper's art. "Two of these birds will attack a cow or llama and kill it with their terrible beaks and claws," says the Rev. J. G. Wood, and, added to this strength and prowess, there is its unparalleled power of flight, which enables it to hunt the preserves of half-a-dozen states, cross vast, wild mountain ranges in search of a new meal, or hang suspended on the watch for prey at a height when even its monstrous expanse of wing is reduced to an almost invisible point. Yet carrion and "a naked savage" bring this monarch amongst birds to grief. They are taken alive by the Mexican Indians and half-breeds in a manner which, though simple in itself, requires both nerve and strength in the trapper. The sole apparatus consists of a newly flayed skin of cow or buffalo. This the Indian places on the ground hair downwards on some bare spot, and then, crawling underneath, turns over on his back and waits. In a short time a condor comes overhead, wheels round and descends on the hide. Immediately his talons touch the skin the Indian seizes the legs, and, starting up, overwhelms the bird and binds him with thongs kept ready; a process, however, which usually meets with a very stubborn resistance. It is just this weakness for rank flesh that is the betrayal of all vulture kind. All through the East it seems as though Nature had kept especially in mind the scavengering duties of these her too hideous children, and meat with that gameyness which is produced by a few days' exposure to a tropical sun is an irresistible attraction to them. The Andes type is no better. The wandering tribes take it by placing a dead horse in an advanced state of unsavouriness within a high wattle enclosure, and noosing the glutted birds when they have fed too freely to rise. And in much the same way, according to Tschudi, in one of the Papuan provinces there exists a deep natural funnel-shaped cavity in the side of a certain valley. This is utilized by the Indian as a ready-made trap for capturing condors. They place a dead horse or mule on the brink of this hollow, and the pecking and tugging of the giant birds presently roll it down the declivity. The birds follow, and being heavy and gorged, are unable to ascend again, clubs and stones finishing off the disgusting revellers to the last one.

Mr. Willard Schultz, writing to the American *Forest and Stream*, gives a curious picture of the superstitions attendant on the procuring of eagle plumes for the head-dresses and robes of "braves." He says: "Another ingenious method of hunting practised by the Blackfeet Indians of North America was the Pis-tsis-tse'-kay for catching eagles. Perhaps of all the articles used for personal adornment eagle feathers were the most highly prized. They were not only used to decorate head-dresses, garments, and shields, but they were held as a standard of value. A few lodges of people in need of eagle feathers would leave the main camp and move up close to the foothills, where eagles are generally more numerous than out on the prairie. Having arrived at a good locality, each man selected a little knoll or hill, and with a stone knife and such other rude implements as he possessed dug a pit in the top of it large enough for him to lie in. Within arm's length of the mouth of the pit he securely pegged a wolf skin to the ground, which had previously been stuffed with grass to make it look as life-like as possible. Then, cutting a slit in its side, he inserted a large piece of tough bull meat and daubed the hair about the slit with blood and liver. In the evening, when all had returned to camp, an eagle dance was held, in which every one participated. Eagle songs were sung, whistles made of eagle wing-bones were blown, and the 'medicine men' prayed earnestly for success. The next morning the men arose before day-light, and smoked two pipes to the sun. Then each one told his wives and all the women of his family not to go out or look out of the lodge until he returned, and not to use an awl or needle at any kind of work, for if they did the eagles would surely scratch him, but to sing the eagle songs and pray for his good success. Then, without eating anything, each man took a human skull and repaired to his pit. Depositing the skull in one end of it, he carefully covered the mouth over with slender willows and grass, and, lying down, pillowed his head on the skull and awaited for the eagles to come. With the rising of the sun came all the little birds, the good-for-nothing birds, the crows, ravens and hawks, but with a long, sharp-pointed stick the watcher deftly poked them off the wolf skin. The ravens were most persistent in trying to perch on the skin, and every time they were poked off would loudly croak. Whenever an eagle was coming the watcher would know it, for all the little birds would fly away, and shortly an eagle would come down with a rush and light on the ground. Often it would sit on the ground for a long time preening its feathers and looking about. During this time the watcher was earnestly praying to the skull and to the sun to give him power to capture the eagle, and all the time his heart was beating so loudly that he thought the bird would surely hear it. At last, when the eagle had perched on the wolf skin and was busily plucking at the tough bull meat, the watcher would cautiously stretch out his hands, and grasping the bird firmly by the feet, quickly bear it down into the cave, where he crushed in its breast with his knee."

In Scotland the eagle, it is said, is often captured alive by a method very similar to those employed in taking its kindred in South America. A circular space, twelve feet in diameter, is enclosed on a spur of the hills haunted by the birds, and a peat wall six feet high built round it, with one small opening at the level of the ground, over which a strong wire noose is suspended. The bait, a dead sheep or lamb, is placed within, and the eagle coming down to it, feeds largely—not wisely, perhaps, but certainly too well—and, like many another of superior creation, feels, after the repast, disinclined for any unnecessary exertion, so casting round for an easy place in the barricade, he espies the low archway, and attempting to leave by it is caught round the neck and killed—at best a poor end for so gallant a bird.

## CHAPTER II.

# *FINCHES.*

### AMONGST CORN AND FRUIT.

In every country in the world, and in all ages, small birds have been conspicuous for good or ill. They have been observed, utilized, petted, and abused in turn by every race under the sun. The Pharaohs owned gilded aviaries on the Nile when history itself was only in bud. Assyrian monarchs had a leaning to "the fancy," and the calm grandeur of Babylonian halls echoed, very probably, the pleasant ditties of caged warblers and cooing of doves. Chinese emperors have amused themselves with the brilliant plumaged finches of their flowery land for innumerable ages; while bird catching and caging is as old as any other institution from the banks of the Ganges to those of the Nile.

Evidence, classical or mythological, of injury done to human industry by these industrious little spoilers is equally old, from the Hitopadésa to Herodotus and downwards.

But it is not with diminutive pillagers in lavender or maroon who "spoilt" Egyptian millet crops two thousand years ago that we have to deal, or with any of their kindred who take toll of rice grains, or feed in endless clouds where bamboo harvests are littering the jungle ground. The page or two I have to devote to them is rather about their comparatively few and for the most part sober relatives of these islands, the sparrows and chaffinches of the stackyards, the bullfinches and cherry-loving thrushes of the orchards.

Legislation has already and wisely confounded the bitterest antagonists of grub-eating small birds by affording them protection during their breeding season from the 1st of April to the end of August, but even this brand new protection may be endangered unless those who are mostly interested exercise a wise spirit of investigation and caution in hearing the carpings of certain critics so remorselessly dissatisfied that surely they will find fault with the municipal arrangements of Paradise if they are ever in a position to speak practically of them. Only the other day an indignant and no doubt well-meaning farmer rose at a local meeting and deduced from a tome of calculations he had made that small birds had in one season eaten grain in England to the value of nearly £770,000. What could be more shocking than such a consideration with wheaten loaves at sixpence the quartern? On the face of it, it would seem to justify her Majesty and her peers, spiritual and temporal, in forthwith ordering the complete and effectual extermination of every thrush or finch in the land. Thus Frederic the Great declared war against the sparrows, because they were too fond of the cherries for which he also had a weakness. The sparrows disappeared, and within two years the cherries followed. Not long ago in one department of France, where every citizen loves *la chasse*, and the small birds find it difficult to hold their own, the loss on wheat from the raids of insects during one twelve months was no less than £160,000. This is the reverse of the matter, and serves to show, if it shows nothing else, how wide are the differences between the contending parties.

In general the happy mean lies between the two extremes. There is a balance in Nature which cannot be kept too clearly in sight. The great Mother knows best the mechanisms of her own establishments. This is why, perhaps, hawks lay but one egg to every two or three the birds they prey on hatch; and why rabbits and mice, the most universally persecuted of rodents, are amazingly prolific.

If legal protection is afforded to grain-eating species,

while sparrowhawk and kestrel, the natural checks on their numbers, are ruthlessly destroyed, it may well happen they become numerous past all indulgence, and an imperative necessity arises to artificially readjust the balance.

Those of our English birds of farm and garden most usually regarded as harmful are as follows, and to their names are attached a few notes on their principal food at different seasons as shown by *post-mortem* examination. Some of the information is taken from Mr. Groom Napier's admirable little work, "The Food, Use, and Beauty, of British Birds," some from my own observation, and the rest from researches of various observers, reports of the Canadian Agricultural Commission, and the like.

The first birds in the usual sequence are—

*The fly-catchers*, of whom nothing but good can be said. All three kinds visiting England, live during the whole twelve months on gnats, "those motes that sting," on hymenopterous insects, and a host of diminutive enemies to cattle and plant life.

*The thrushes*, coming next, some six species in all, are not so unquestionably innocent.

*The missel thrush*, relies during December, January, February, on holly and mistletoe berries, on haws, earthworms, slugs, snails and anything of the nature he can pick up. This is varied all through the summer by many caterpillars and a little garden fruit, especially gooseberries. In the autumn he has to return to wild berries, and is keen on snails and slugs.

*Fieldfares and redwings* are not here long enough to do any mischief. They pillage the hawthorn hedges and ivy bushes of Nature's alms, and take a certain number of snails, etc.

*Song thrushes* have been well abused, nor are we prepared to say the abuse is undeserved. They are unquestionably fond of fruit, currants being their chief delight; but work energetically in our behalf at all other seasons. A friend of the bird thinks "the thrush, like the blackbird, is doubtless extremely useful in moderation, when its numbers are in proportion to the extent of farm or garden ground. When they are very numerous, however, they are induced to feed upon fruit; but our experiences tend to show that they prefer insects and mollusca to fruit. The same remarks which apply to the thrush apply also to the blackbird."

Everywhere but in the orchard the mavis is useful; even there a remembrance of the pecks of slugs he has gormandized, the great earthworms he has drawn from the ground and shaken to death by the hundred, like any terrier, and those piles of snail shells round his favourite anvil stone, should stay the destroyer's hand.

*Blackbirds* are not quite so black as chance and fruit growers have painted them. All I have said of the former bird applies to this. In the kitchen garden they are invaluable, if fruit is netted from them a short time before it is plucked.

*The ring ouzel* is an inhabitant of the wilderness, where it enjoys unlimited small snails and the insect life of the uplands.

*Warblers.*—That subdivision, known scientifically as the *Sylvidæ*, contains a numerous host of unimpeachable friends of the agriculturist or gardener; and friends, moreover, which by a rough system of reasoning he values. The *hedge sparrow* feeds under the hedges all the year round on seeds of weeds and small insects; *robins* take, perhaps, a small quantum of currants to vary their animal dietary, but they may safely be left in the protection of legend and favouritism.

*Whinchats, wheatears*, and the like, if not amongst "the unco' guid," are useful in their way, while *whitethroat* and *wood wren* destroy plenty of noxious insects.

*Titmice.*—Over these dainty little pinches of feathers the battle of the birds has waged long and hotly. They have been accused of stripping trees of buds (and especially fruit buds) in a reckless and wanton manner. But in nearly every case it has turned out that the buds pulled off were already the home of a larva, which would effectually have prevented their arriving at maturity. Their natural food is to be found on trees and amongst herbage, and consists of all those multitudinous insects that, if allowed to multiply unchecked, would devastate our crops and wither up our flowers.

"Let, therefore, the titmouse be permitted to follow its avocation as it chooses, and to range the fruit trees, fields, and gardens unchecked. For, in truth, the little bird is working with all its might in our behalf, and is attacking our worst pests at their very root and source. Its microscopical eye discovers the eggs of noxious insects which have been deposited in spots where they will find plentiful nourishment when they are hatched, and in half-a-dozen pecks it will destroy the whole future brood. The eggs of the terrible leaf-roller caterpillar, so tiny but so destructive, are devoured in vast numbers, as are those of that plentiful nuisance, the little ermine moth," writes the Rev. J. G. Wood, and we can fully endorse what he says.

*Wagtails*, the *Motacillidæ* of naturalists, do good service in thinning the swarms of summer insects; we doubt, in fact, whether any one has ever called their usefulness in question, while their ways are dainty and their gracefulness conspicuous.

*Larks.*—Against skylarks stands the indictment of scratching newly sown grain out of the soil, and the little excavations made for this purpose are often to be seen during the spring months. Wheat or barley properly drilled in, we should fancy, would be far beyond their reach. Nor is it difficult to argue in their favour that even a chance of feeding thus must extend over a very limited period. At nesting time, when many mouths have to be fed, grain of all sorts is out of reach, and resource must be had to the abundant and ever present harvest of seeds from weeds, wireworms, insects, etc.

*The chaffinch* feeds "in January and February on seeds,

grains, and berries; in March on seeds and insects; in April on seeds, green food, and insects; in May on seeds and insects. Almost all finches that live on seeds and berries feed their young principally on insects," writes Mr. Groom Napier. "In June the chaffinch feeds on insects, berries, and fruits; in July and August the same, with the addition of a little more seed; in September, October, and November on seeds, berries of many sorts, and grain." During these autumn months it haunts stackyards with flights of sparrows and searches for scattered grain. It will descend in flocks amongst newly sown turnip seed, and does, undoubtedly, a good deal of mischief there.

In allusion to the frequent notices of the formidable gooseberry grub in the columns of *The Field*, that excellent observer, Mr. Doubleday, of Epping, observes that a brood of young chaffinches will soon clear a gooseberry bush from these grubs. It is, therefore, the obvious interest of gardeners to protect and encourage chaffinches in the breeding season, instead of taking so much trouble to destroy them or frighten them away. It must be admitted that this beautiful and most cheerful spring songster helps himself to our radish seed as soon as it has germinated; but, without attempting to palliate this species of petty larceny, may we not regard its services in destroying the gooseberry grub as a full equivalent?

*The greenfinch.*—This bird is fond of seeds, and has an extraordinary and insatiable appetite. His value, or the reverse, to British agriculturists is not very clearly defined.

*The goldfinch* is not numerous enough to be of much economic consideration. One peculiarly good point he has, namely, a passion for downy seeds of any sort. This was a happy thought of Nature's, and the love of the goldfinch for the pernicious thistle (or rather its wind-scattered seeds) and the like, suggests him as being as useful in regard to numbers as he is unquestionably handsome.

*The bulfinch* strips our cherry trees in a very lawless

manner of their buds, and is consequently persecuted by fruit growers. Fortunately he is not numerous, nor is he difficult to scare away, when this milder treatment is adopted.

*Starlings* are unquestionably useful. They scour our meadow lands, effecting as they go a wonderful clearance of wireworm and the detestable "daddy-long-legs" in all its stages. Amongst cattle, and even riding on the backs of sheep, they are still useful, having a taste for the parasites of such animals. On marsh lands they feed largely upon small mollusca, worms, etc. Occasionally a raid is made upon cherries, but there is no other indictment to be brought against them.

*The swallows* are worthy of our fullest friendship, I think most people will allow. Leaving out of consideration the facts they are the symbol of summer, and typify the very poetry of motion, their existence is spent in keeping within bounds the myriads of winged insects, which might otherwise overwhelm us as Pharaoh was overwhelmed when he had refused for a fourth time to set free the Israelites!

*The sparrow*, it will be noticed, we have reserved for the last. The antiquity of his transgressions is beyond dispute. Perhaps he fell firstly with the prince of the nether world himself. In the most remote Egyptian hieroglyphics he is represented as then old in iniquities, bearing a name, *sa-me-di*, signifying "bird of destruction," and an outline on tomb and obelisk indicating death and scarcity. This is a point for his opponents which they have overlooked. His credentials have been faulty from the beginning, his passport has never been signed by the lords of creation; and the farmer of to-day, in offering a reward for his head, is only inheriting a long and classic feud!

It is true the sparrow does not seem to care much for his disrepute and outlawry. He is equally cheerful "on the house tops" as rusticated. I doubt if he was happier, guided by the ribbons Aphrodite held and fed [on gilded seeds of Asphodel, than he is now, sharing the swine's breakfast and dining on a dunghill. There is a story telling how sparrows were nearly exterminated in Germany by a heavy premium paid for their heads which enlisted the enthusiasm of every *knabe.* In Norway and Sweden, too, for one reason or another, I noticed some time ago that sparrows were almost absent from homestead and stubble; but in the main *Fringilla domestica* would seem to thrive on persecution. "It would be a pity," thinks one tender-hearted ornithologist, "if the sparrow were completely extinct." I must say there seems but little prospect of this. Only a few months since seven thousand heads were capitated for by one club in one English shire, "and yet there seemed to be but little difference in the number of birds about," plaintively observes the Judge Jefferies of that ornithological Star Chamber. No doubt in such cases as this there *is* a difference, but other sparrows come in from neighbouring districts and fill up vacancies.

The transgressions of sparrows are many. They eat corn, they shell peas, they spoil fruit, they encourage plumbers by building in ill-chosen places, they bully martins and swallows (a serious offence), and monopolize their nests; straw is drawn from thatched roofs, crocus, as well as other flowers, are pulled to pieces, better birds are driven away and much mess made. The indictment is heavy, and, what is worse, I fear a true bill must be returned in every case. I say this reluctantly, for I love the sparrow's pleasant chirrup as he basks in the first sunshine of the spring, and have seen in him every trait of love, anger, vanity, cunning, and resource that the bird world can produce. He is an epitome, in grey and brown, of natural uncultivated life.

As for his actual food it is infinitely various. One "Monograph of the Sparrow," recently published, puts it down as corn, green or yellow, and nothing but corn; but this is foolish prejudice. Mr. Groom Napier makes it more various: "January, February—seeds, grains, refuse, insects; March—green tops, seeds; April—insects, green tops; May—

larvæ, seeds, green tops; June, July—fruits of the garden, seeds, insects; August—grain, insects, berries; September—grain, berries; October, November, December—grain, refuse, seeds, berries."

"The proprietors of gardens have a special reason for gratitude towards the sparrow. Gooseberries are a favourite fruit, whether fresh or preserved, and we are too often doomed to see our trees lose their leaves, and the crop of fruit fail, solely through the attacks of the gooseberry-fly, the dark grey grubs of which are so plentiful and voracious. These grubs are very pleasing to the sparrow's palate—though, by the way, it seems rather strange that a bird should have any particular sense of taste, considering the formation of its mouth and the substances on which it feeds—and accordingly are killed in great numbers by that indefatigable bird. For many successive days the sparrows may be seen filling their beaks with gooseberry grubs, and bearing them off to their young.

"The wireworm, again—a pest that is perhaps more universally dreaded than any other of the insect tribes—is a favourite food of the sparrow; and it has been well calculated that, though the sparrow is said to eat a bushel of corn annually, it saves a quarter by its depredations among the insects. The sparrow, in fact, has recourse to that most effectual system for ridding the plants of the destructive insects which, when performed by man, is termed 'hand-picking,' but which cannot be achieved by man with one hundredth part of the success that attends the bird."

The sparrow hates cats. When the poultry are whistled together at feeding times, numerous small birds join the dinner party. Pussy then creeps up and hides herself amongst the hungry group, by this time quite used to her tactics. Watching her opportunity, she suddenly darts upon her victim, which she stealthily carries off in her mouth, returning warily again to the charge. Taking advantage of this, the most effective way to scare birds from fruit trees is this: From two pegs fixed in the ground stretch a piece of wire, then procure a cat or kitten three parts grown, put a leather collar on it, and attach it to the wire by a slip knot, also of wire, so that the animal can at will range the whole length of the pegs. The presence of the cat, combined with the rattling of the wire at its every movement, have proved a capital protection against the feathered marauders. For this to answer properly, however, the trees should be in rows, as in the case referred to, and the pegs fixed at the extremities, the wire thus running parallel to them.

The sparrow, in fact, needs to be kept in bounds rather more than any other bird.

The whole matter is one in which caution and reasoning are especially necessary, since there are side issues and cross-bearings on every point. The purely insectiverous birds, for instance, might be thought, like Cæsar's wife, to be above suspicion; yet it could be shown that, by eating a thousand forms of life that prey on more injurious insects, they are doing very dangerous labour, and many other instances could be given. One thing only is certain, the majority of birds do us yeomen-service, however much some few may transgress, and any tampering with the often ridiculed but nevertheless essential "balance of nature" is a matter deserving the gravest and most serious consideration on all sides.

## By Stack and Stubble.

While there can be no doubt we have lost, and are losing, some of our larger indigenous birds—the eagles, the kites, the bustards, the ravens, choughs, and such like—there seems, on the other hand, little recorded diminution among the smaller feathered *fauna* of copse and hedgerow, in spite of the unreasoning warfare just alluded to. We still have the nightingale, "that sovereign of song" that Spenser loved, the sparrows King Alfred fed, the "throstle, with his note so true," who sung to Shakespeare in pleasant Avon woodlands; and, in fact, nearly every one of those lesser birds enshrined in poet's verses, or enbalmed in our rich and historic folk lore, with which song or story has made us familiar. Perhaps the finches of the underwood and the wild birds of marsh and mountain top owe their immunity from extinction to their shyness and retiring habits. The whitethroats might be as scarce as bitterns were they equally noticeable; but, as matters stand, who cares to molest the former—that delicate little fragment of drab and cream-coloured feathers that hunts in the nettle forests and hides its grass-built nest amongst densest tangles of briar and bramble? We might have obliterated the ouzels, again, as we have the auks, had they been half so valuable for food or so dull-witted as the gare-fowl. This, and much more of the same kind, goes to show that when left to their own devices Nature very rarely suffers any species of bird or beast to be "wiped off the slate." It is only when man, the lord and bully of creation, comes upon the scene that the balance is disturbed; races and species going down before his insatiable appetites and endless vanities. It was not Nature, for instance, who did away with the amiable but heavy dodo; it was South Sea whalers, and all for the poor reason of sharpening their sailor's knives upon the stones his gizzard contained. The birds of paradise are dying to deck the dresses of savage tribes, and humming-birds to fringe fans and glitter on fair but thoughtless heads. Penguin flesh was very good eating the cods-men of the North Seas knew, and the fact was ruin to the species; and just so the buffalo is being recklessly converted into glue and pelts for portmanteaus, until we are within measurable distance of his extermination; and the price of elephants and elephant ivory going up every day, as they become scarcer and scarcer in their Indian or African jungles.

Nature retaliates, it might seem, by multiplying unduly some smaller birds and beasts, not to mention lesser insect plagues. But leaving locusts and larva out of the question, even the naturalist must recognize sometimes that certain manner of birds or beasts are unduly redundant. There is the rabbit in Australia, for instance, working shocking havoc on the sheep runs, and living in a very Arcadia where stoats or weasels are unknown, and ruining biped and quadruped with its ceaseless fecundity. The sparrow in America is as bad, and the Senate has arraigned, condemned, and excommunicated him several times, without, however, any perceptible effect on his cheerfulness or numbers. We forbear to enlarge upon the devices prepared for the beguilement of this little scourge of Christendom, as his enemies call him, since the erratic propensities of the sparrow not only lead him to trespass on every man's land, but bring him sooner or later into every man's trap. For this reason, and the fact of his small mercantile value, few lures are devoted to his special circumvention. Of those that are, however, the "bat-folding-net" is one of the most destructive. This consists of two twelve-foot bamboos, slightly bent and joined at their thinner ends, having a net of small mesh stretched between them nearly down to the lower or handle ends, where the net is turned back for a foot or so to form a trough-like pouch. When in use one man holds the lower ends of the bamboos, and applies the net, spread between them to ivy on walls or trees, haystacks, eaves, etc., and wherever the birds may be sleeping at that hour of the evening; while another man with lantern and stick beats the foliage, etc., and the affrighted birds dash from their roosts to meet the wall of net, falling after a brief struggle into the open pouch below.

Barring these perky little finches that Venus loved, we have in this country few kinds of birds that assemble in great flocks, and can thus be killed wholesale either in revenge for fancied injuries done or for "the pot." Abroad it is otherwise. In Germany, for instance, they are overrun with starlings. On November evenings the fowlers of the Upper Rhine watch for the arrival of the great flights of starlings. A little cloud is seen on the horizon, which gradually approaches and grows into a black spiral column, which at last almost darkens the air and deafens the ears with the chirping of its innumerable host of birds. After a few spiral turns they suddenly perch in a body on the trees and reeds, which appear laden with leaves and fruit, and bend under their weight. The fowlers mark the spot where they settle, and then set up an immense curtain of nets on poles in an advantageous position, and so contrived that they shall fall when a cord is pulled. This done, they leave the chattering throng to settle down into their roosting-places, while they themselves go home to supper. At midnight, however, they return, and posting themselves round the roosting-place of the birds, suddenly raise a tremendous shout, and with long sticks and stones drive the frightened birds towards the net. The whole flock rises *en masse* and makes for the net, which, as soon as they beat against it, is pulled down, and the whole flock enclosed. They are left to be strangled in the meshes or drowned in the marsh till daylight, when the fowlers again return, to take them out and dexterously twist the necks of those which are not dead already. Sometimes as many as ten thousand are caught at one fall of the net, but not more than five or six thousand are taken, the others being allowed to escape, for fear of glutting the market. They are taken to Strasbourg and sold at the rate of 3*d*. to 4*d*. per dozen.

There can be little doubt that though this may be a good speculation for those immediately concerned, it is a ruinously bad one for the Rhine lands at large. "Perhaps there is no bird that does so much good to the husbandman as the starling," says Swaysland. He is the terror of every sort of grass or corn devouring grub and pupa. The inquisitor of the meadows, he believes in summary jurisdiction, and the wireworm or grub hauled into his presence must expect very little mercy from that *beak*. Sometimes they come to be regarded as a nuisance, or available ingredients for a pie in our own southern shires. A correspondent writes: "As owner of a larch plantation of over one hundred acres in

Somerset, I can give the following plan that I used to adopt some twenty-five years since, when my larch trees were young, by which plan I caught hundreds of birds, including starlings, fieldfares, and other similar birds, on any dark, still night. One man carrying, say, four sheep bells, one man with a lantern, and another a long light stick, one sheepdog, enter the plantation after seven o'clock, the first man shaking the sheep bells, which drowns all sound of footsteps; the second man turns the light on the trees, when the birds can be seen, apparently stupefied; the third man knocks them down; the sheepdog retrieves them. This may be called poaching; but where the birds roost in thousands they may be used as food, and certainly are excellent eating."

And another of these, we must think, ill-advised landowners suggests we should have some openings made by stripping the trees in two or three places right and left through our plantations so as to admit of many clap-nets, and then send a person to quietly beat the birds towards the nets, when we shall capture a score or two, as starlings do not rise and fly away, but flutter along the branches.

These birds migrate, unobtrusively but widely, though the fact is not generally recognized. There can be little doubt the greater part of those flocks seen on our marshes and downs during the winter have come from Norway and Sweden.

Though starling *pâté* may seem a poor substitute for pigeon pie, the truth is, nearly all small birds are more or less good food. Nothing could seem less appropriate for this purpose than the swallow tribe; yet Buffon tells us swallows roost at the close of summer in great numbers on alders by the banks of southern rivers, and are taken in vast quantities to be eaten in some countries, as Spain and Silesia; and again we read, "The martins grow very fat in autumn, and are then very good to eat. They are taken very largely at Alsace, in nets;" adding that these birds, like all the swallow tribe, are excellent for the table when young and fat.

The Spaniards—who eat all sorts of "little game," in season and out of season, with no regard for plumage or habits—capture bee-eaters and rollers at night, by going round and pouring water into holes in banks and trees where they roost, at the same time holding a net over the entrances, into which the affrighted birds speedily dash. When out on these expeditions, both Little and Scops owl are frequently captured in the same way, or even with the hand, owls and rollers alike appearing strung up above the stalls of the next day's market-place.

Birds in Spain are taken when roosting on the ground by parties of two, the one carrying the bag and also a bell, which he tinkles monotonously, whilst the other carries a light; the idea being that the bird supposes it is only some vagrant bell-wether, and remains till the captor with the light puts his hand upon it. My belief in the usefulness of the bell is limited; that of the light is an established fact. Yet the bell is used in this manner in many countries. In Somersetshire and Andalusia we have noted its use. The Lincolnshire fenmen employ a bell when netting plovers; and the lark, another very edible and marketable bird, is betrayed by its sound in France. The method is disgusting in its unvarnished brutality. A dark night being chosen, two men are required. One has a bell which he constantly jingles in one hand, and a lantern in the other, with which he throws a light along the furrows of the newly turned corn-lands where the quarry roosts. The other, who goes ahead, has a stick, at the end of which is a short strap of heavy leather, and a sack. When a bird is seen cowering under the light it is approached cautiously, and a single stroke from the leather "flap" extinguishes its life without spoiling it for to-morrow's market. The professional manner of catching larks is by means of a trammel net. This is about thirty-six yards long and eight wide. At each end of the net there is a pole, and the lower edge is weighted so as to drag along the ground. Men holding the poles and raising the front of the net tramp forward. If they are lucky all the birds at roost on the ground covered will be taken, the net being lowered to the ground whenever captives are felt or heard to rise against the meshes. Moonlight is fatal to the sport, and wet nights equally so, for then the net is too heavy to drag. It is an improvement if the men holding the end poles each lead a horse by the bridle, as his footsteps—to which the birds are accustomed—drown theirs; or the men sometimes ride the horses, as we have seen represented in old prints. In the winter, when the snow lightly covers the ground, larks may be taken in considerable numbers by horsehair nooses. This is accomplished by driving pieces of wood into the ground so that some three inches are above the surface, and they should be about three yards apart; then, after the fashion of a laundress's clothes-line, stretch twine from stump to stump; now make nooses in lengths of horsehair, and suspend them from each line, so that the running loops dangle freely, about two inches from the surface of the ground; scatter black oats about the noose, and larks, in seeking to pick it up, will find themselves held captive by the horsehairs. Clever though these designs are, gourmands might sigh in vain for larks on toast, were it not for the clap-net—that deadly device in skilful hands. Two nets, twelve yards long (and, when open, covering the ground twenty feet wide), are neatly laid down upon the ground. It is impossible, without a diagram, to explain the rough, but very effective, machinery by which a pull of the rope held by the birdcatcher will make those harmless-looking nets spring into the air, and catch the birds, either on the wing, or on the ground. The nets act so quickly, that the eye can scarcely follow their spring. Anything on the wing crossing them four feet high will be shut in instantly. It is better to catch the bird before he has time to settle; if he touches the net with his feet, he is off instantly.

The next process is to put out the "brace bird." This bird always wears his brace, with a swivel attached, consisting of a piece of string made into a kind of double halter, and put over the bird's head, and the wings and legs are passed through, the feathers falling over, and rendering it invisible. The brace bird is then put on his "flur-stick;" this is a straight stick, which, by means of a hinge at its lower end, is made to rise and fall at the will of the bird-catcher by means of a string.

Then, when any bird is seen coming, the flur-stick is gently pulled up, the brace bird all the while standing on the stick is made to hover with his wings and show himself. This, of course, is to attract the wild birds to the place, which purpose is also attained by "call birds" put out round the net in cages, whose notes, especially when there are others of their kind in the neighbourhood, attract great numbers. Thus, no doubt, are procured those melancholy festoons of Nature's choristers we see in the gamedealers' doorways. Personally, we think that good as this little bird may be at table, *aux trufles*, legislation should sternly proscribe his presence there, or even his entombment alive in any of the cruel little cages with which some of us associate him. He should be as sacred to us music-loving nations of the West as doves were to the Greeks or the Ibis to Egyptians. This same "seraph of the sylvan choir" is a bird of strong passions, and often stirred by love or hate. The fowler, with the gross practicalness of his kind, knows this, and takes a mean advantage. If the season suggests the predomination of the gentler sentiment, then a female decoy, whose wings are tied and a lime twig placed over her is used. The male in paying his court thus gets hopelessly entangled.

But if there is a note of challenge in the song we hear coming from under the clouds, "then," says a learned fancier, "start at break of day, carrying with you a well-trained singing lark. Tie its wings, so that it can do no more than hop about the ground, and under the string slip the ends of two lengths of flexible whalebone, the projecting ends of which must be well smeared with bird-lime, and cross each other over the decoy's back. Watch where a lark rises, and put down your bird near the spot, the wild bird will drop like a stone on the back of the trespasser, and it is caught by the lime."

One more method of taking these diminutive wild-fowl—a curious and sportsmanlike method we might almost say—if it does not take us to the end of our available notes, will at least probably exhaust a reader's patience: "To-day some bird-catchers brought a number of pipits for sale," writes an Indian traveller. "The method of capture was ingenious. Sheltering themselves under a screen of leaves, they would creep to within about thirty feet of where the birds were running about. They then push forward a series of bamboos, which fit into one another like the joints of a fishing rod, the top one being provided with a pronged twig smeared with bird-lime. This, on coming in contact with the bird, of course holds it fast, until the native runs up and wrings its neck 'in the name of Allah the Compassionate!'"

Small birds as food are much more popular amongst other races than amongst the Anglo-Saxon. Every continental market-place is at times an ornithological exhibition. Under the olive-groves of the Ægean Islands, and all through the Mediterranean, finches and warblers at all times of the year are liable to get themselves into nets or toils of varying make.

Just outside Port Said I have seen something novel in the way of bird-catching. Two Arabs, with casting-nets, were walking along the canal bank, here dotted with patches of scrub a foot or eighteen inches high. Marking down some unfortunate small bird, they stalked and cast their nets over the bush on which it had taken shelter, seldom making a bad shot, though their "bag" could not have been a heavy one, as none of their victims were larger than a titlark; several were the tiny fantail-warbler (*cisticola*), so plentiful throughout Egypt, particularly on scrubby ground anywhere near water.

The wheatear is almost the only other edible small bird we recognize in these islands. Gilbert White, it will be remembered, remarks, they "appear at the tables of all the gentry around Brighton and Tunbridge who entertain with any degree of elegance;" and elsewhere we read, "It's favourite haunts in this country are the South Downs, and in the neighbourhood of Brighton, Lewes, and Eastbourne great numbers are taken in traps, which are set on the downs cut out in the turf. The habits of the birds in running to shelter on the least alarm are considered in the nature of the snares set for them, which are made after this fashion: Pieces of the turf are taken up in solid masses, and propped up over the holes from which they are cut; thus a sort of hollow chamber is formed, holes are left at the opposite end of the space formed beneath the turfy cover, and in the hollow itself nooses are set vertically, supported on small sticks; the birds rushing in for shelter are caught by their necks in the nooses, and fall an easy prey to the setter of the trap. Quantities of wheatears are thus taken and sent to the different markets, where they realize from 9*d.* to 1*s.* 6*d.* each. Their price has very considerably increased of late years; from 6*d.* to 1*s.* a dozen used to be given formerly in a plentiful season. Then the shepherds on the downs were the chief trap makers, capturing sometimes from fifty to sixty dozen in a day, and a custom then prevailed of people visiting the traps, taking out the birds (if there were any caught), and leaving a penny in the trap as a reward for the shepherd—a somewhat primitive method of proceeding which would not hold good at the present time."

The late Frank Buckland declares the best trap for wheatears is the common nightingale trap baited with a meal worm.

But for our part we think there is a very good time ahead for the small birds, and probably an enlightened public opinion will learn to recognize in them faithful allies on the farm lands, and delightful associates in the uplands and wildernesses.

Abroad they take a very practical interest in their small game, especially in the French mainland, as also in the Mediterranean islands. From Corsica, for instance, vast quantities of birds are sent to the Gallic markets, and they are indeed the most popular "game" in the island.

### "Among the Corsican Scrub."

We, that is to say, W—— and myself, on one occasion had finished supper, and were smoking, in grim discontent, over a roaring fire of fir-cones in a little Corsican inn, the howling north wind rattling the badly joined window frames, and the rain pelting on the glass like so much small shot, as indeed it had done with scarcely a pause for seven days, everything feeling dull and uncomfortable, even a few feet from the blaze, when a footfall sounded in the passage, and the next moment our door was thrown open by a much bewrapped Frenchman, who immediately advanced with outspread hands, giving us a tremendously cordial greeting after the fashion of his country, and without more delay than served to divest monsieur of his two wet overcoats and uncoil a dozen yards of "comforter" from his neck, we refilled our pipes and plunged into the subject that so much interested us.

Monsieur R—— was our chief reliance for sport in the island, whither we had come to spend the winter. We had made his acquaintance on board the French steamer, and as he was a well-known chasseur, he had promised to show us whatever sport there was to be had in Corsica, hence his welcome appearance on the wet evening of which I write.

"Could we not get a moufflon?" asked my comrade. But R—— shook his head. It was out of the question in such weather as this.

"What then?" we said, somewhat anxiously. Forthwith, our guest propounded an idea he had formed that we should have a rough day in the macchie and river estuaries, after (and W—— heard it with a blush!) the very small game in which continental sportsmen delight, varied by perhaps a duck or two,—in fact, anything we came across, until such time as the clouds chose to lift from the hills and give us a chance of searching their summits for better game. This was the best he had to offer us. Though not much, it was better than hanging about the hotel verandah, smoking indifferent tobacco, and wondering where on earth the sunshine we had come so far to find had got to. It was therefore agreed on, and an early start the next morning being arranged, we said good night, and "turned in," in a much better frame of mind.

Half-past eight a.m., and the light clatter of wooden shoes on the red tiles outside my room roused one even before the *fille-de-chambre's* tap on the door, and the ostentatious clatter of her hot water can became audible. A little while later, we two Englishmen met in the coffee-room, where we were soon joined by R——, who pointed out the happy fact that it was a glorious morning, with a lovely sky, and every prospect of fine sport before us.

Breakfast over (and on such occasions one is apt to make short work of it), our mules were announced at the door. We, therefore, strap up the game bags (which R——, to whom we left the provisioning of the expedition, has filled so full of lunch and bottles that they can only be fastened with the greatest difficulty), and when this is over, lighting our pipes, we sally out to our steeds in the courtyard, ready saddled, their head-gear bedecked with numbers of little red tassels which they shake to keep off the flies. My two companions, who have beasts of discretion, mount without trouble, but mine is of a different mould, and wheels this way, and that, taking "snips" with his teeth at the trousers of the by-standers, and discharging sundry kicks that enlarged the circle of spectators with remarkable quickness. So I wait till he settles down for a minute, then rush into close quarters, and before he can move a leg, I am safely "on board" with every intention of staying there. Then away we go, our two men, with the guns and a couple of dogs, following behind as fast as they may, our steeds cantering along down the narrow village street, scattering the old women and children on every side, and creating a vast panic amongst the long-necked chickens.

Once we get clear of the little Corsican capital the blue Gulf of Ajaccio bursts on us, brilliant as a sapphire fresh clipped from its mother rock; here and there are feluccas stealing about its calm surface with long white sails—fishing perhaps, or off to the coral grounds at the head of the gulf. On both sides of the lovely bay the land slopes upwards, terraced with dark-foliaged lemon groves, or left unreclaimed in the wild dominion of prickly pears and cactus, giving the hill-sides a strangely mottled appearance, as cultivation and Nature thus struggle side by side. Far away to the north-west, where the blue water ends, Monte Rotondo rears high over the valleys and plateaus, its head still crowned with heavy snows, the remnants of last winter's storms. Not only was the view fine, but the air was delightful after the rain, and the bright sun overhead seemed to put new life into the small birds along the roadside, and I could not help lingering behind the others, occasionally, as the road turned about amongst glorious gardens of orange trees, every twig of the forest of dark-leaved trees heavy with green or golden fruit, each leaf and blade wet with dew and rain that flashed in a hundred colours as the sunlight glanced down from above. An orange garden has always been a wonderful sight to me!

Half-an-hour's riding brought us to a branch road, down

which we plunged and pulled up at an old ruined chapel, shaded by a large olive tree. Here begins our shooting ground, so we shoulder our cartridge bags, load up the guns, and, leaving one man in charge of the lunch, set off with the other and the dogs for the open *macchie*, or the close-leaved and densely planted shrubberies of wild myrtle, arbutus, and leutiscus that clothe nearly all the higher ground in Corsica with a delightful canopy of evergreen verdure. Amongst the various sweet berries of these shrubs astonishing hordes of blackbirds and thrushes revel all day. We put them up on all sides, to the great satisfaction of our French companion, who began peppering away at the *petit gibier*, and we, with a little hesitation, followed suit. It was pretty enough shooting, however unorthodox. An infinite variety of brisk little birds rose from the irregular growth of arbutus, and with a couple of flicks of their wings were over the bushes and out of shot in an extraordinarily short space of time. Nothing but the quickest of snap-shooting was possible, and our light guns, and special small loads of powder and shot, had to be very "straight" to keep up a creditable average. W——, the deadly on grouse, scored several misses when the fun began; of course I did no better; while R—— led us up the rises, fusilading as he went, as though we were storming a Russian battery!

Where the arbutus berries were thickest a perfect cascade of small birds, thrushes, blackbirds, and pipits rose on every side. "No wonder there is so little *game* in the country," said my companion, looking at me ruefully as he began his third score of cartridges, "if much of this sort of thing goes on!" But I pointed out to him it was only an experiment, as I much wanted to know where and how the French markets were supplied with their small birds, and he sighed and bowled over two thrushes right and left.

A modification of this process is practised in the Ionian Islands, and a correspondent has penned a pleasant account of it, which I cannot resist reproducing.

"Far different is the course adopted in the Greek Islands, for so soon as the middle of October arrives, may you expect vast flights of thrushes, with which are mingled a few of the missel thrush (called here on the principle of everything large coming from Africa, the Barbary thrush). When it is fully ascertained that these birds have been seen in numbers, which is always the case by the 20th of October, then every one is bitten with the desire to go into the olive-groves to 'whistle for thrushes.' As this is rather a curious proceeding, and opens up a new phase in thrush character, I cannot do better, perhaps, than describe a morning expedition in one of the Ionian Isles, on which occasion I was inducted into the ceremonies. It was towards the end of October that I started for the fern-covered, woodcock-haunted glades of Gorino, in company with a Greek gentleman skilled in 'bird murder.' How well I remember how gloriously the morning dawned, the early grey shadows softening the harsh outlines of the forts under whose guns we passed, ere winding up the steep hill upon which the picturesque little village of Potamo is placed. From this elevated spot the view was magnificent; far away below us lay numberless olive groves, over the tops of whose trees could be seen the grey still waters of the harbour, and the shores of the Emarantine Island now gilded here and there with the awakening beams of the sun, which was driving the vapour in clouds from the bosom of the sea. Salvador's high crest yet wreathed in mists; its sombre slopes clothed with the ever verdant holly and ilex, while it seemed yet summer, so calm and warm was the air, its silence unbroken save by the mournful whistle of the curlew on the sandbars below, or the harsh chattering notes of the wary jay in the thick trees above us; around and about were mossy little dells thickly clothed with high bushes of myrtle and laurel, the velvet sward around luxuriating in the dew that our hasty passage brushed from off the brown tangle of herbage which served as shelter for the shy woodcock. On we journeyed, through ravines, past hill-sides where the crimson fruit of the arbutus, called here 'Frooli di Montagna,' or mountain strawberries, tempted us to linger awhile, past vineyards where the sere and rapidly dying leaves augured little as yet for that purple cluster which would depend from every branch when the heat of summer had again clothed them with verdure, past the orange trees and their now small unripe fruit hiding amid glossy dark green leaves, until some miles had been traversed; and we stood at last, before the sun had risen high enough to dispel all the night mists on the far-off mountains, on the summit of a hill overlooking the sea, from which we expected the thrushes to arrive. We were not the only tenants of the spot we had selected, however, as there were two or three countrymen stationed under the cover of as many trees. My friend now produced his whistle, which was a round hollow piece of silver (though mostly constructed of copper) about one inch in diameter, convex on one side, and concave on the other, with a hole right through the centre. The concave part is placed in the mouth, pressing against the teeth, and by inspiring the breath, and modulating the tones with the closed or open hands, as the case may be, a very perfect imitation of the song thrush's note is the result. This the arriving or newly arrived birds hear, and imagining that it proceeds from the throat of one of their species, alight in the trees which surround and conceal the treacherous imitator, and quickly fall a prey to the ready gun. So infatuated are they, that enormous quantities are killed by this method early in the season; in fact, I know one person who shot one hundred and four, besides other birds, to his own gun in one day.

"In this particular instance the effect was wonderful, for the whistles had not been sounding long before high up in the clear air, some half mile away over the sea, some tiny specks appeared. 'Thrushes?' queries my friend of another posted a few yards from him. This ascertained, the whistling proceeds more vigorously than ever. The voyagers near us, they appear now to waver in their flight, and hover together in the air; this indecision is, however, overcome by a few persuasive notes from the call, and they descend into the trees with an undulating sweep. Theirs, alas! is no happy welcome to a foreign shore. Bang! bang! go the guns almost simultaneously, and five or six lay on the velvet turf; the rest take to flight, but are followed and nearly all shot in detail, for while the fatal whistle sounds they may be approached, with a moderate degree of caution, and will sit with their heads on one side and their bright eyes peering into the underwood, until the shooter gets almost as close as he likes to them."

To return to our personal adventures. When we had shot enough small birds for a good store of pies, we got monsieur to come on to the borders of an overgrown wilderness of tall bamboo-like reeds, forming a dense jungle of many acres in extent at the estuary of a small river flowing into the gulf. Here we turned in the wild Corsican dogs, and got ourselves ready for whatever sport the fates might send us, W—— going round to the far side, while the other two guns stayed on this. The first thing to rise was a duck, which R—— promptly "potted" at fifteen yards' distance, and retrieved in person with a very fair imitation of an Indian war-whoop. Two other ducks were put up from the thicket of waving stems, and we heard W—— get off both barrels, as the birds went over to his side. Then came a pause, owing to the dogs having struck work and disappeared, to be found after a quarter-of-an-hour's whistling a couple of hundred yards back, busy lunching on the remains of a dead horse. Of course they were "reproved," and then we started again, but the walking was very poor, at one time all bog or mud reeking, as we leapt from one spongy tussock to another, with foul malarial taints, again sand like that of the sea-shore, or worse still, a vast desert of rounded pebbles such as continental rivers are fond of depositing when they get a chance. However, we trudged on this varied surface, getting in the first hour about a dozen shots at ducks, of which only seven were successful, owing to the birds hardly giving us a chance in the thick cover, and then the reeds gave out, and our forces met where the lagoon narrowed up to the mountain torrent that had given it rise. Here we rested for a moment to fill the pipe of peace, but this necessary operation was hardly done when the sharp ears of the Corsican guide caught the cry of some partridges higher up, and though likely to be "red legs" and great runners, we set off after them at once, getting two as they rose from under the side of a rock, the others—if there were others—making good their retreat to the nearest strong cover. Forthwith W——'s enthusiasm for partridges rose to a high point, in which I backed him up, for the lovely sweet-scented macchie was much superior to the marsh below; so we changed our duck-shot cartridges for smaller shot, and marched into the red legs' territory.

A lovely shooting ground it was—not particularly easy to work, but delightful from an æsthetic point of view. Noble hill-sides gleaming and warm under the bright Mediterranean sun, dotted about with clumps of olive and oak, over which the kites and hawks swept in circles, frightening out—as the shadow of their wings passed along—whole herds of small birds from the deep foliage of the myrtles and arbutus. Gardens of orange and peaches, just coming into flower, luxuriated on the warm southern terraces; here and there the white walls of a farm-house peeping out from amongst the verdure or the little peaked roof of a wayside chapel, in which the image of a saint standing under a ceiling of blue, spangled with golden stars, called on the passer-by to drop on his knees and breathe a prayer. Amid this charming hunting ground we strayed all the morning, taking things rather too easily for making much of a bag, but picking up a hare, three or four partridges, and a brace of quails out of a bevy of which we ought to have got more; but we were not on the look-out when they suddenly rose and dodged round a rock with their pretty chirruping cry, affording us only a very quick snap shot.

Then we lunched under a wide spreading cork tree, with the blue Gulf of Ajaccio extended far and wide from the low ground at our feet, and the pale snow-fields of the Corsican Alps glittering at our backs. Thanks to the care of monsieur, who had prepared it, the meal was only too complete, and he now presided, beaming over the array of everything the hungry sportsman could desire: fascinating pies of myrtle-fed songsters and cold game from the hotel chief's larder to eat, while for drinking there was the bottled beer of the Saxon, and the light wine of the Gaul, honey stored by upland bees, smelling of mountain pastures, and brought down from far inland by peasants, who had also supplied their goats' milk cream for us to eat it with; and when all these dainties had been disposed of there came a glass of Chartreuse to wind up with. Truly a Frenchman understands the science of eating. Such a lunch, "though it might be magnificent, was not war," or rather shooting, and, need I add, that when it was over we smoked a pipe or two with great deliberation, and then coming to the somewhat tame conclusion that we had done nearly enough shooting for the day, contented ourselves with strolling homewards along the beach, getting a couple more ducks and three or four hares from a stony bit of half-reclaimed land that bordered the sea shore.

There is not much to be said for Corsican sport. To make bags of any size it is necessary to go very far inland, where the best shooting is found. As to the famous moufflon, or wild sheep of the island, I have been after them once or twice, but it is much to be feared their day is near its setting, as they are well nigh extinct.

Thrush hunting here in our own country is regarded as a fit amusement only for country bumpkins, or at most a pastime for Master Tommy home for his Christmas holidays, and revelling in the delights of a new gun—a pleasant alternative for him, perhaps, from hunting cats in the shrubbery with his sister and the terriers; but abroad the matter is different. In Italy and Spain the orange groves and olive wastes are depopulated of useful small birds, as we have seen, and Gould, in his "Birds of Great Britain," gives a graphic account of "La Tenderie" in Belgium. "The thrush is a great source of amusement to the middle and of profit to the lower classes during its autumnal migration. Many families of Liége, Luxemburg, Luneberg, Narum, parts of Hainault and Brabant, choose this season for their period of relaxation from business, and devote themselves to the taking of this bird with horse-hair springes. The shopkeeper of Liége and Verviers, whose house in the town is the model of comfort and cleanliness, resorts with his wife and children to one or two rooms in a miserable country village to enjoy the sport he has been preparing for with their help during the long evenings of the preceding winter, in the course of which he has made as many as from five thousand to ten thousand horse-hair springes, and prepared as many pieces of flexible wood rather thicker than a swan quill, in and on which to hang them. He hires what he calls his Tenderie, being from four to five acres of underwood about three to five years old, pays some thirty shillings for permission to place his springes, and his greatest ambition is to retain for several years the same Tenderie and the same lodging, which he improves in comfort from year to year. The springes being made, and the season of migration near, he goes for a day to his intended place of sojourn, and cuts as many twigs, about eighteen inches in length, as he intends hanging springes on. There are two methods of hanging them: in one the twig is bent into the form of the figure 6, the tail end running through a slit cut in the upper part of the twig. The other method is to sharpen a twig at both ends, and insert the points into a grower, or stem of underwood, thus forming a bow, of which the stem forms the string below the springe, and hanging from the lower part of the bow is placed a small branch, with three or four berries of the mountain-ash (there called *sobier*); this is fixed to the bow by inserting the stalk into a slit in the wood. The hirer of a new Tenderie three or four acres in extent is obliged to make zigzag footpaths through it, to cut away the boughs which obstruct them, and even to hoe and keep them clean. Having thus prepared himself, he purchases one or two bushels of the berries of the mountain-ash with the stalks to which they grew, and which are picked for the purpose after they are red, but before they are ripe, to prevent their falling off; these he lays out on a table in the loft or attic. The collection of these berries is a regular trade, and the demand for them is so great that, although planted expressly by the side of the roads in the Ardennes, they have been sold as high as £2 the bushel; but the general price is five francs. We will now suppose our thrush-catcher arrived at his lodgings in the country, that he has had his footpath cleared by the aid of a labourer, and that he is off for his first day's sport. He is provided with a basket, one compartment of which holds his twigs, bent or straight, another his berries; his springes being already attached to the twigs, he very rapidly drives his knife into a lateral branch and fixes them, taking care that the springe hangs neatly in the middle of the bow, and that the lower part of the springe is about three fingers breadth from the bottom; by this arrangement the bird, alighting on the lower side of the bow, and bending his neck to reach the berries below him, places his head in the noose, and finding himself obstructed in his movements, attempts to fly away; but the treacherous noose tightens round his throat, and he is found by the sportsman hanging by the neck, a victim of misplaced confidence.

"The workman, who at this season earns a second harvest by this pursuit, carries on his industry in wilder districts, or he frequently obtains permission from his employer to set springes in his master's woods. In this case, he supplies the family with birds, which are highly appreciated as a delicacy, especially when almost covered with butter, with a few juniper berries, and some bacon cut into small dice and baked in a pan; the rest of his take he sells at from 5*d*. to 10*d*. per dozen.

"No person who has not lived in the country can imagine the excitement among all classes when the Grieves arrive. If the morning be foggy, it is a good day for Grieves; if bright, bad Tenderie! The reason is obvious: when the birds arrive in a fog, they settle at once in the woods; if bright, they fly about seeking the most propitious place for food. I may observe a singular feeling of honour is engendered by this pursuit. Nobody will think of injuring his neighbour's Tenderie; a sportsman would carefully avoid deranging the springes. If, when shooting in your own covers, a few are taken for the table, you would hang a franc piece conspicuously in an empty springe for every dozen birds taken. The law is very severe on poachers who place a springe on the ground to take partridges, woodcocks, or snipes; but if three feet above ground, the law says nothing, and save as a trespasser, the placer of springes in the trees of a wood not his own property would not be punishable. The number taken is prodigious—as many as one hundred and fifty thrushes have been found executed in a Tenderie in one morning. The younger members of families of the highest ranks commonly follow this amusement before a gun is placed in their hands.

"It may be readily imagined that before five thousand springes are set in a Tenderie of four or five acres, a fortnight or three weeks will have elapsed, even should the grocer, the linendraper, or publican, be assisted by his wife and children. The amusement is common to all the family—wife, boys, and girls. Many a small tradesman eats little else during his vacation at his Tenderie besides Grieves and Buem. From Liege to Tilf, thence to Ayvale on the rivers Meuse, Outhe, and the Amblere to Chauspritaine on the Vesdre, where the rivers are for miles shut in by precipitous banks, covered with low woods, scarcely an acre is unlet for Tenderie during the months of August, September, October, and November. The first fortnight of August is occupied in preparations, the rest of the time is the harvest of Grieves."

## CHAPTER III.

# *CROWS.*

### AMONGST THE ROOKS.

THERE could not well be a thinner excuse than that which justifies the shooter's intentions as he goes out at the season of new green leaves to ravage the homes of his ancestral servitors the rooks. He says, perhaps, as he fills his pockets with cartridges, something about the need of adjusting the balance of Nature, and of the damage the young "crows," already noisy in the avenue outside, will do presently to the spring corn. Ten days ago had you asked him, his opinions were all in favour of the dusky birds, and he recognized that their plumage is but a physical chance, and not the livery of sin some have pretended. And a fortnight hence he will acknowledge that they do yeoman service on grass and plough, searching with restless inquisitiveness for grub and wireworm, and giving all and sundry of these and such other small but powerful enemies of the farmer the shortest shift. Yet for the brief period intervening between the feathering of the young birds and their incorporation with the wandering flocks of their parents, squire and farmer are remorseless, and persecute them with a vigour not a little remarkable. But very likely the fact that this is a chance of burning powder coming after an abstinence and before another spell of the sportsman's Ramadam, accounts for the change of principle. Moreover there is delight simply in being out of doors in "the leafy month of June."

Rooks have a peculiar aptitude for selecting for their home a spot of dignity and beauty. They are always associated with stateliness and repose. No one ever found their nests in a disreputable spot—such as a gooseberry bush for instance, where we have known a magpie to build—among the stony curls of a heroic statue like ribald jackdaws, or even among chimney stacks with the storks. Just as engravers give a little "local colour" to an Indian etching by bringing in a palm or two, and accentuate Arabian sands by a camel in the background, so an English artist never finishes up his cathedral precincts or surroundings of a ruined manse without throwing in the nucleus of a rookery and a bird or so coming home with sunset. No doubt these birds have built in the plane trees of Cheapside, where, by the way, kites built only a hundred years ago, in Gray's Inn Gardens, and in a few such other places, but this does not spoil the argument. Where we find them most numerous and available for sport is in the avenues leading to lordly mansions throughout the shires, and in the great elms that the foresight of our ancestor planted behind grange and castle to keep off the north wind, and to shame, perhaps, shallow, sceptical descendants, who live as if their lives marked the bounds of time, and who, cutting down, plant nothing for those who come after.

There are countless traditions regarding the cunning and feudal instincts of the rook. No money-lender ever had a greater interest in the succession of great estates than these sable retainers of long-settled families. One authority tells us gravely they will desert a rookery that is about to change human ownership, and that a tenantless mansion where familiar faces have once been they abhor. Foresters more prosaically aver they can tell when an elm has the wet rot even sooner than the woodpecker, their distant relative. To bark their trees will drive them away, and so may a ring of paint round the bole, as surely as though with human eyes they associated that fatal mark with axes and woodmen.

But where some subtle auguries of inconvenience, of which we bipeds cannot fathom the origin, do not frighten them off, they are very tenacious of their homes. All through the winter of their discontent, when the early barley is as snug under the frost-hardened ground as gold in a usurer's chest, and grubs of every kind are at a premium, they keep an eye upon the wigwams that swing in the wind over the bare avenue, and a little later on, when the elms are thick with their unacknowledged copper-tinted inflorescence, they hold a curious festival in the tops—perhaps an "at home," suggested by matronly forethought, "to bring the young people together,"—when the whole clan reassembles for a day or two, and "small and earlies" are held with vivacity and success. Then nests are overhauled and even added to—a spectacle that prompts the wandering stranger to write to his favourite paper, pointing out that the winter must surely be one of the mildest on record. But those who live among rooks know that nothing comes then of this freak. In April they set to work in earnest, industry and jealousy reigning supreme in the colony; faggot upon faggot of sticks is fetched and crossed over last year's foundations, tufts of wool and the like are gleaned from sheep-walks and pastures; and the last touch is put to the structure by an egg—three or four perhaps—no doubt in the opinion of each enamoured couple the most delightfully shaped, the most delicately blue-tinted, and the most artistically mottled of any in the park.

But we have almost forgot to shoot our "branchers" in the interest of the steps leading to their hatching. The rook battue is the most popular form of this sport. The squire asks his friends down to the number of a dozen or so, according to the number of trees and nests, and for a day, or perhaps two days, the fun is fast and furious. The happy time to hit upon is just when the "squatters" are venturing upon their trial flights. Were they younger they might keep to their nests, where it is barbarism to shoot them; and were they older, then the shooting would come to a speedy termination by the whole colony migrating with natural expeditiousness to less disturbed regions. As it is, some of the stronger birds go out to the pasture oaks, and we have to go after them, wading for a shot waist deep through wet, sweet-scented meadow parsley, or deep swathes of grass almost ready for the scythe, before we come back with our trophies, as likely as not wet through. But what seems to our selfishness the choicest sport is to be alone this early summer weather with our trusted little rifle only for a companion, and license to be as unsociable as we will. Then we can lie at leisure on the wide blue carpet of the wood hyacinths, or, sauntering down the drives, come unobserved upon many a curious bit of nature, and witness many a little comedy or tragedy of the woods that the powder-burners up at the hall never dream of. In this way we have spent many a summer morning, lying perhaps concealed among the green commas of the unwinding bracken and the thin covering of the new leaves, while the rooks fed their young ones on the low trees about us, all unsuspecting of our presence.

Within the limits of the crow species, as we know them in England, are included some birds very dissimilar in outward garb, though there is a perceptible family likeness amongst them in character and outline. Their physical blackness is but the reflex of the character they bear amongst the less thoughtful, marking them as outlaws by flood and forest, common enemies, excommunicated beyond hope of redemption, whom it is virtuous to slay and witty to revile!

I am not going to white-wash them, but suggest the latest views of other country-side observers, and my own, on the depth of their negritude. It is useless to pretend human observation can detect a track of shame or remorse in crow kind for even the most palpable and flagrant offences brought home to them. Nest-pillaging village boys they detest, and keepers, when they have a gun with them, they respect;

but for the rest of humanity they have an undisguised contempt.

The jackdaw of Rheims was a false bird to the extent of his contrition for the theft committed. Had he been a daw true to his breeding and colour, as far at least as mundane probabilities go, he would have defied and derided the Lord Cardinal's "holy anger," and cared not a *sous* for the plenary absolution. Crows of all kinds are strong in their self-conceit, though this is best seen abroad amongst the white collared birds of the Transvaal or the slim-built *Corvus splendens* of the tropics. Here, at home, the crows (with the exception, perhaps, of the rook) shun civilization, keeping much to themselves; nor is it to be wondered at, for constant trapping and shooting is making every one of our six or seven species scarcer each season.

How can the raven thrive, for instance, when shepherds proclaim he tears the eyes from lambing sheep, and keepers swear he spits in pure wantonness every kind of young animal upon that remorseless black pionard, his beak! No need to describe his geographical distribution. He is a citizen of the world. "His sable plumage reflects the burning sun of the equator, and his shadow falls upon the region of perpetual snow; he alights on the jutting peaks of lofty mountains, and haunts the centre of vast untrodden plains; his hoarse cry startles the depth of the dense primeval forest, and echoes amongst the rocks of lonely islands of the ocean: no *ultima thule* is *terra incognita* to him; arctic and antarctic are both alike the home of the corbie crow." Johnson, the African traveller, found him, pied in colour by the way, when he was fighting and sketching on lonely Killamanjaro in middle Africa, and a raven was the last fresh meat Lieutenant Greely and his starving Americans tasted when they wintered under the bitter crags of Cape Sabine within the arctic ice.

As far as England is concerned these birds have been driven into the fastnesses of the north, the Welsh hills and some such wild localities as the Yorkshire scars or Cumberland wolds. There is little to be said for their protection or encouragement; any little good they may do as eaters of carrion or destroyers of useless lower life, is lost in the immensity of their tenantless feeding grounds, while, on the other hand, they undoubtedly tyrannize over game and weakly sheep. "They will pursue even the buzzard, the goshawk, or the eagle, to endeavour to obtain from him his own capture," writes the Rev. F. O. Morris, and consequently it may be understood they would not hesitate to attack a mountain hare far from cover in the snow, or rend a young sheep astray from its companions. The only facts commending this sable bird of Thor to our care is his place in history and legend, and the tender heart of the naturalist, which is Buddist in its encircling indulgence. Choughs and jackdaws are equally neutral in character, the former—crows with scarlet legs and bills—keep to a few rocky headlands round the Cornish or Yorkshire coasts. It is long since they were seen in any numbers east of the Solent, though Shakespeare knew them well enough, and recently one observer writes from Dover:

"The chough has not been seen about these cliffs for many years. About twenty-five years ago I saw one from the parapet at Archcliffe Fort, on which I was leaning, looking seawards at a lot of gulls. It was flying amongst the latter, and came within ten yards of me, so that I could see its orange bill and legs. A local naturalist has just told me that he saw a chough near the South Foreland some twenty years since. I think the jackdaws, which swarm in these cliffs, occupying every available hole, would drive the chough away."

Jackdaws, on the other hand, are well known wherever there are escarpments or ruins. No one can be familiar with the south coast without recalling its jackdaws. In spring I have seen them quarrelling and building amongst the yellow wall flowers and Valerian on the ledges of the white cliffs; sweeping out in clamorous schools at every real or fancied danger—the measured tread of a coastguard above or the shadow of a gliding kestrel crossing their nursery floors; and in summer they curvet with their young over the breezy downs, or descend upon the cliff crofters' potato plots, but no harm is committed there or elsewhere by them.

With infinite disgust have I met town gunners turning out of an afternoon to harry this cheerful and harmless little bird amongst his breeding places in the ruins, and in particular one such party comes especially prominently to my memory. I was walking down "Tweed side" and passed under the ruins of Drochil Castle, once owned, it is said, when Scotland was an independent monarchy, by a noble baron who turned his restless genius to the invention of the guillotine; and subsequently, under direction of his sovereign, illustrated the working of the affair on his own person with the assistance of a few regal retainers! This stronghold was overgrown with ivy, and abounded in jackdaws who cawed and chivied one another through casement or port holes, adding life and interest to the scene. I sat down and thought how well their presence befitted quiet. "Surely no voice in Nature was ever more suggestive of long undisturbed repose, more significant of the statelier forms of peace, or more in harmony with old baronial possessions than the pleasant clamour of the jackdaws up amongst the chimneys and turrets. Not only do they enhance the tranquillity of the ancient castle, but they add a solemnity to the minster; the poets are quite wrong when they say the 'steeple-loving jackdaws' note is dismal.' Down the strath, when I had left the birds, with my heart full of friendliness to them, I met three or four townsmen armed with cheap breechloaders, about whose errand I speculated for a time. It was only when retracing my steps the same evening up the glen the wretched mystery was explained. Those gentlemen of clothyard and scales had had a field day amongst the birds, the castle was silent and deserted, and along both sides of the approach were some sixty or seventy greydaws, dead, and impaled in reckless mockery on the points of a hurdle fence at distances of ten yards apart, a most melancholy avenue under the rays of a rising moon! It is hard to draw a hard and fast distinction between what is cruelty to animals and what is not; but there ought to be no difficulty in morally defining wanton slaughter or distinguishing it from legitimate sport.

In coming to the rook we come to a very fertile source of controversy which would fill a portly volume if argued out to the bitter end.

"Rooks do endless damage to seed corn," say the farmers, "and moreover peck holes in root crops, thereby letting in the frost, thus ruining acres of keep at a time when it is most valuable."

"Besides this," suggests velveteens, who only knows some half a dozen birds, classing all the rest as "vermin," "they carry off plenty of young game in the season, and play havoc amongst the c-o-ops if left unguarded for any time."

Of these accusations, the first is undoubtedly the most serious. Though the bareness at the base of their bills is not due, as has been ingeniously suggested, to constant friction with the soil, yet they are unquestionably great and successful diggers. If wheat in a dry March is put in lightly or broadcasted, the rooks will find it out and undoubtedly take their toll. Yet there is a cheap and easy remedy at hand which solves their delinquencies at once, and makes us safe, moreover, from small birds.

There will be no further need of bird keepers if farmers would adopt the following process: Take one pint of gas tar to two gallons of warm water, for eight bushels of corn, and well mix in the same way wheat is dressed for smut. When sown neither rooks or game of any kind will disturb it, nor will the dressing injure the growth of the seed.

They by no means depend on one class of food for support. A close observer illustrates this. He says:

"During the last few years I have brought up young

rooks by hand and turned them loose in the garden, where they have been continually under my observation, and I have never yet seen one of these birds eat a slug or a snail.

"Earthworms they will drag up and eat by morsels; earwigs, beetles, chrysalises and flies they will swallow whole in any numbers. When given to them they will eat cold potatoes, cheese, biscuits and eggs, raw or cooked, and game.

"One spends a great part of his time waylaying sparrows. When caught he holds the sparrow down with his claws, while he plucks it, regardless of its shrieks; he then pulls off the head, and, after eating the body, buries the head and intestines. One of my rooks once caught a large frog, which he tried to swallow whole, but one leg protruded from his beak and was immediately snapped off by his fellow bird. While gardening we have frequently offered numerous slugs and snails to the rooks; but, seeing that they never touched them we, of course, now destroy these garden pests as soon as discovered.

"My rooks are quaint and amusing pets, easily tamed and very intelligent."

About their sagacity there cannot be the slightest doubt. They are rarely caught in traps, though later on I give some ingenious devices for that purpose; sticks driven into the ground and connected by simple zigzags of string will keep them away from any place. They have a horror of any sort of beguilement, nearly as great as their repugnance to a gun. The farmer who can get near enough to the rooks unearthing his corn to shoot one of their number, will not be troubled by the survivors for some time to come. The difficulty is to get within range. I failed so often in the attempt that at last I fell back upon a Snider rifle; with this I have several times got to within one hundred and fifty yards of a feeding flock, and a shot "into the brown" or rather *black*, has caused a ridiculous panic without, however, any great harm being done to the birds.

Sometimes their attention is transferred from corn to

meadow land, which latter they "scarify" after a day or two's work as though a patent harrow had been once or twice over it. Bad as this looks, it hides a good purpose. The rook does not feed on grass, nor has he time for mischief pure and simple. He has been indulging in wire-worm and cockchafer grub—dainties of which he is very fond—and the amount of these wretched, ruinous grubs a flock will make away with in a morning's campaign is simply astonishing. Let the farmer run his light roller over the well-probed leas and bless the rooks, they are not the least useful of his feathered allies. Perhaps the game-keeper can hardly be invited to say so much. Here, for instance, is a sad story from a writer in *The Field.*

"My keeper one morning observed about half a dozen rooks engaged amongst the coops of young pheasants, and, suspecting their object, drove them off. The next morning, having fed and watered the young birds, he went to his cottage, and, looking out about six o'clock, saw a strong detachment of rooks from a neighbouring colony in great excitement amongst the pens. He ran down, a distance of two hundred yards, as fast as possible, but before he arrived they succeeded in killing, and for the most part carrying off, from forty to fifty birds, two or three weeks old. As he came amongst them they flew up in all directions, their beaks full of the spoil. The dead birds not carried away had all of their heads pulled off, and most of their legs and wings torn from the body. I have long known that rooks destroy partridges' nests and eat the eggs when short of other food, but have never known a raid of this description. I attribute it to the excessive drought, which has so starved the birds by depriving them of their natural insect food that that they are driven to depredation. It will be necessary to be on guard for some time; bad habits once acquired (as with man-eating tigers) may last even more than one season. Probably the half dozen rooks first seen amongst the coops tasted two or three, and finding them eatable, brought their friends in numbers the next morning. In former years, when drought has prevailed, instances have been recorded of rooks robbing nests of the callow brood; and in the winter, too, when the ground has been too hard for them to get food, they have been known to hawk after and kill small birds."

But rooks afford some legitimate sport in May time, and such transgressions as these are very rare indeed, the result unquestionably of being very greatly pressed by hunger. Probably not one keeper in fifty has lost birds by rooks. Crows (and crows, I may point out to the unlearned in country side lore, are quite distinct from rooks) *do* do some damage to our pheasants and partridges. In Norway and Sweden they and the magpies have obliterated ryper and grouse from the fell sides. Here at home they cater for their young with an atrocious want of discrimination generally bringing prompt vengeance upon them. Only let us be certain when the luckless corbie is arraigned and executed that we have got hold of the real criminal.

A suggestive story in point should make many a game-keeper of conscience look aside as he passes his museum on barn door or ash tree.

"Some time ago there were several letters in *The Field* regarding hedgehogs eating eggs. Within a single season there have been two distinct cases come under observation, that have conclusively settled the question for ever. The first is this: I had a tame duck laying under some tops of trees that had been recently felled in the wood where I reside. There were five eggs in the nest. On the following morning there were only two and a piece of shell. On the following night I put down a common rabbit trap at the nest, let into the ground, and covered over. About ten p.m. I heard something crying out (similar to the noise made by a hare when in distress). Upon my going I found a very large hedgehog in the trap. I took it out, killed it, and set the trap again. About eleven p.m. there was another large hedgehog in the wood pile, which I killed, and set the trap again. I went again the next morning at five a.m., and found another large hedgehog in the same gin, making three hedgehogs in one night caught at the duck's nest. Since then the duck has been sitting in the same nest undisturbed by anything. The second case occurred recently. One of my men came to me with a face as long as a fiddle. 'Master,' says he, 'the crows have been and spoilt a pheasant's nest that you knew of down the wood, by the withy bed.' I asked him if he was sure it was crows. 'Come and see for yourself,' was the answer. I went, and sure enough there were nine eggs destroyed out of fifteen. They appeared to have been bitten half through. It then came to my mind about the hedgehogs eating the duck's eggs, and I was determined to find out and prove what it was destroying these eggs. I took the remaining six eggs home, and inserted a very small quantity of strychnine into each egg, and sealed them up again, and took them back to the nest where the others were destroyed. The next morning the man and I went to see if anything was there, when we found an immense hedgehog flat on his belly, and very much swelled up, not a yard from the nest, and quite dead, and as if in the act of crawling away from the nest. Only two of the eggs were partially eaten. Is not this conclusive evidence that the hedgehog is a great enemy to the pheasant and partridge?" And, I may add, evidence that crows and other birds often suffer for guilt not their own.

I might enlarge to any extent on jays and magpies, those picturesque brigands of the coppices. As imported by its specific name, Morris observes, the acorn is the most choice *morceau* of the jay, and for them he even searches under the snow; but he also feeds on more delicate fruits such as peas and cherries, as well as on beechmast, nuts, and berries, corn, worms, cockchaffers, and other insects, larvæ, frogs and other reptiles, even mice, and is deterred by no qualms or scruples in making away with young birds. These birds, in the autumn, are said to hide away food for future use, under leaves in some secure place, and in holes of trees. They are great egg-suckers also, and the nest of the missel thrush, song thrush, and blackbird suffer greatly, both when they have eggs and young ones. In the latter case many a furious fight I have witnessed, and the gallant conduct and boldness these birds exhibited in defence of their helpless brood was truly astonishing, since they pursue the jay with unrelenting fury, in and out of thickets where it would try to gain shelter. Occasionally I have observed it succeed if there were only one pair of birds defending; but it often happens that other pairs come to their assistance whose nests or young ones are in the immediate neighbourhood, and these, boldly and unitedly concerting together against a common enemy, often drive it ignominiously away. The magpie is a little better in service to humanity.

Of these two, as of all the rest of the genus, I can only say that in place and in reason they are a distinct gain to our allies by covert and meadow. When they trespass they trespass badly, worse indeed than the majority of birds; but of this I am certain, that all the crow kind within our four seas do less harm to agriculture in the aggregate than a single shower when the hay is down, or corn is ripe; and much less harm to game than a thunderstorm (or an inch of snow on the high grounds), when grouse or partridge chicks have grown too big and bulky to shelter under their mother's wings.

## Crows and their Capture.

A reasonable and philosophical view must indeed soon be taken of the work done for mankind by the crow, the rook, and their kindred. Were it otherwise, we should hesitate before divulging any of those many and cunning secrets devised for their destruction which a store of human enemies have scattered through the pages of "Manuals" and "Treatises."

For, heretical as it may sound, we have a strong feeling of friendship for the dusky brotherhood. Perhaps it will be suggested we have never suffered materially at their hand, or we might be less indulgent. We do not allow this, for we have felt, and bitterly resented at the time, nearly every form of indignity to which the corvine species can put either the sportsman, the naturalist, or the farmer. And yet there appears, in our mind, no legitimate need to consign the whole race to that hideous barbarity "the gamekeeper's tree," for when their numbers are moderate the good they do, and the life infused into often desolate regions, far outweigh their transgressions. At least, this is the writer's experience, an experience, moreover, practical and not altogether inextensive.

The raven, for instance, the first of his kind in size, strength, and cunning, if a hundred fables about him are to be believed, has been my companion in many a lonely ramble up Highland passes and over the seldom trodden wastes of the wild western coast. Perhaps he does occasionally take a juicy young grouse, when he fancies a change of diet would be useful, and young hares or mountain rabbits playing about far from cover undeniably suggest dinner to him. But the harm done in this way is small, even when all carefully recorded, and we could write off with very little grudging in our game books each season a few brace of birds or fur to his account. Then, there is the crow—a bird of evil omen all over the world, which, nevertheless, contrives to live a happy and useful life from the verge of perpetual ice along the Nova Zembla shores to the rim of the antarctic circle. The oldest Vedas tell us how he fell from Paradise, and the most ancient Cinghalese writings record his original sin: "In wrath for their tale-bearing—for had they not carried abroad the secrets of the councils of the gods?—Indra hurled them down through the hundred storeys of his heaven;" and the Pratyasatka adds that "nothing can improve a crow." In India he is the common enemy; kites knock him off the roof ridge or wrench away the meat he has stolen from the kitchen; pariah dogs come on him round the corner and shortly dine on "black game." Has the "mem-sahib" lost a silk handkerchief? Then it must be that —— rascally crow on the cotton tree who has taken it, and forthwith the "butlerwallah" attaches an appetizing bit of meat to a couple of feet of string and throws it into the compound. At the other end of the string is a small stone, and when the bird of sable plumage swoops down and flies off exultantly with the morsel in his beak, the string very speedily swings itself round his body, and the result is an ignominious tumble to earth, and an inglorious scuffle on the sand till native fingers loosen the tangle. But then begins the worst part of this proceeding for *Corvus splendens*. He is taken into the shed, which goes by the name of kitchen in India, and plucked remorselessly, being divorced from every vestige of plumage, as he struggles and kicks between the butler's knees; and then, in this plight—truly a sorry one—he is released, to die of melancholia, we should fancy, on the nearest roof top, for a live plucked crow, "naked and ashamed," is about the most woe-begone spectacle in ornithology that can be imagined.

Native children are also proficient in capturing the much-abused crow. A lively and strong bird is obtained, little used to such indignities, and is pegged down to the ground in the open on his back with forked twigs, which are driven in over his wing-bones. He very speedily lets the whole neighbourhood hear of his misfortune, and the wild birds, flocking round him, crowd so close that at length one is seized in the sufferer's claws, and convulsively held until the fowler rushes from his ambush, and secures it for himself.

Then, again, comes the bitter part, for the birds Apollo loved are taken, and after suffering numberless indignities at the hands of their small tormentors, each receive a daub of cobbler's wax on its beak, between the eyes, and in this three or four red or green feathers are stuck—a style of borrowed plumage comical in the extreme—the unfortunate birds being a laughing-stock, not only to their biped enemies on the ground, but to their friends amongst the boughs, if one may judge by the clamour with which they are received aloft.

Shakespeare speaks of the crow as "ribald;" Prior, "foreboding;" Dyer, "lurking;" Churchill and Gay, "strutting;" Dryden, "dastard;" and so on. In every land they have a bad name. "Yet they do not wear their colour with humility, or even common decency. On the contrary, they swagger in it, pretending they chose that exact shade for themselves. . . . In the verandahs, they parade the reverend sable which they disgrace; sleek as Chadband, wily as Pecksniff. Their step is grave, and they seem ever on the point of quoting scripture, while their eyes are wandering towards carnal matters. Like Stiggins, they keep a sharp look out for tea-time, and hanker after flesh-pots." Hiawatha knew of a land of dead crow men, and in Thibet the wandering pilgrims say there is an evil city of crows. Those who have dipped into northern fiction must remember the Swedish "Place of crows and devils," and the Norwegian "Hill of Bad Spirits," where the souls of wicked men fly about in the likeness of the same unfortunate bird. In the latter country the crow is an undoubted nuisance, and a terrible poacher; so he is shot and trapped without mercy, but (as usual) seems to thrive on persecution. The "crow pen" is a common sight in the Scandinavian backwood villages. It is nothing but a huge birdcage, but formed of slender saplings seven or eight feet in length, and about four feet high above the ground. The spaces on top are left pretty wide open for a time, until the crows have become used to going in and out to get the bait—and there are few places where their kind will not go for that purpose—and then when familiarity has bred hardihood the top pieces are put close together, leaving only a hole through which the crows can easily drop with closed wings, but too small for them to flap through when they rise from the inside. Great is the rejoicing of the farmer and his small children, and prodigious the clamour of the birds on the outside of the trap, when this cage is thronged with an entrapped multitude.

The rook is such a pleasant neighbour in a country house that he is generally and properly protected, but occasionally he is "wanted" either as a misdoer, guilty of agricultural offences, or as a victim to the modern falconer, who finds in him a convenient quarry to enter haggards upon. Mr. J. E. Harting tells us that rooks are taken for this purpose in two ways. The first is to get a boy to climb a tree in the rookery, taking with him a long line with a noose at one end. The noose must be carefully adjusted over the nest in such a manner that when a rook has settled down to roost in the evening, the falconer on pulling the other end of the string at the foot of the tree may catch it round the legs. It is in this way that herons are generally caught for the same purpose. A certain amount of care must be exercised to ensure the line running freely, and also to ensure getting the bird down nicely. The other method is to set traps behind a plough, and to get the ploughman to shift them from time to time as he proceeds. The trap need not be large or heavy, and a short line or peg will prevent a rook flying away with it. If the spring be too strong or the teeth too sharp, the jaws may be bound with list so as to prevent a risk of breaking the bird's leg. In putting on the list, of course care must be taken that it does not impede the closing of the trap; otherwise the rook on springing it would extricate his foot and get away.

But, says another friend of the birds—"In his industry the farmer has but few such friends, or the insect world such foes. Up in the morning, before the dew is off the grass, the rooks are hard at work disposing of that 'first worm' which proverbially falls to the lot of the early bird.

Like detectives, they are perpetually on the watch to arrest some one, and woe to the insect, grub, or beetle whose evil ways are discovered. There is no appeal from a rook. It holds its sessions where it chooses, and they may look for summary procedure who come under this admirable bird."

This only makes all the more ungenerous the device of the Wiltshire husbandmen. Though we have no experience of the success of the method, it is said rooks are taken by them as follows: A number of cones are made of dark-coloured paper. At the bottom of each of these is placed some corn, and round the upper edge is smeared a little birdlime. The cones are then stuck about a field, point downwards, where the rooks resort, and, on their coming there, they observe the corn, and thrusting their heads in to obtain it, the cones become stuck to them, rendering them blind, and they may be captured in that state by hand. In folk-lore they hold an honourable place. They are said to connect themselves with the fortunes of families, deserting their elms when disaster overtakes the house; and Cosmo di Medici, visiting England two centuries ago, was especially struck by the pride the peerage took in its rookeries. "For these birds," said he, "are of good omen."

The jay is a crow with the men of science, in spite of its gay dress, and lets out the secret in voice and inquisitive ness. Though "the brigands and tyrants of the coppice," they are one of the few birds of brilliant plumage native to England, and do but little harm to the game of our woodlands, it is on the small birds that they chiefly wage war. Their clanship and the interest each takes in its neighbours' concerns is very remarkable. A writer in a long-extinct journal gives a very amusing account of the way in which this trait in the jay's character is turned to use for his destruction. Describing an orchard in German Alsace, he says: "It was pretty extensive, covering, I should say, a couple of acres, and its trees, which were, *all but one*, in excellent trim, were chiefly apple and cherry trees. The lonely one which made thus an exception to the rule was truly in as desperate a plight as any fruit tree could ever be. Leafless, barkless, broken-down, and bare, it was but the very ghost or skeleton, at best, of an old apple tree. At its foot was a small hut, about the size of a Newfoundland's kennel. This hut was made roughly, of a few willow branches, with both ends stuck in the ground, tunnel-shape, and covered over with a few handfuls of evergreens. Karl, my active German guide, who had brought a large clasp-knife, thereupon proceeded to cut down a few more branches and leaves from the nearest hedge, and he interlaced these with the framework of the hut, so as to make its interior tolerably secure from the prying glances of the jays on the morrow.

"The next morning, early, 'we were all there,' as the Americans say. A cool morning it was, with a fresh breeze blowing, and the dew yet on every blade of grass, when we left the keeper's house. Karl was carrying a large pan of bird-lime, a bundle of small branches about a foot long, a long stick notched at one end, a large and long empty cage, and a smaller one containing a live jay. On his back was strapped a small bundle of hay. When we reached the hut, he, first of all, thrust the hay inside, and placed the cage out of harm's way in the hut. Then he cut the string which held all his bits of stick together, and taking them, one by one, he thrust each of them separately in the notch of his long stick, dipped them, turn by turn, in the fresh lime, and fixed them here and there on the uppermost branches of the apple tree, wherever any forks in the branches allowed a resting-place for them.

"When he had arranged the lot, the sun was getting pretty well up in the sky, and we crawled into the hut. Our position there was not remarkably comfortable, but the prospect in store was rather cheering, and that made a compensation for the somewhat cramped posture we were for the time being forced to adopt.

"Meanwhile, by peering through the leaves I saw a band

of jays coming at a very slow rate across the tops of the forest trees towards us.

"'Now the fun will begin,' whispered the *garde* in my ear, and his eyes twinkled merrily. Saying this, he placed and held the little cage in front of his knees, and began poking his fingers through the bars.

"Screech! screech! screech!" at once shouted the captive jay, at the same time attacking vigorously the keeper's hand, and all the while keeping up incessantly its extraordinary clatter.

"'Screech! screech! screech!" replied immediately the astonished wild jays, pausing at first in their surprise, and settling on the branches nearest to them to listen.

"'Screech! screech!' pursued the tame bird, and the others, wondering no doubt what on earth was being done to their *confrère*, set sail without further parley, and drew nearer to try to find, doubtless, where the aggravating assault was being committed. In a few moments, the trees around us were covered with them, turning their big heads this way and that way, making their eyes sparkle and shine like beads, and showing themselves off, oh, so beautifully! with their wonderfully bright plumage, amidst the ripe cherries and the green leaves which surrounded them. It was almost a pity to disturb and catch them, but the keeper did not see it 'in the same light.' 'Screech! screech! screech!' exploded Karl's bird under his manipulations; and, lo! whilst I was watching one of the strangers in his evolutions something fell behind me with a great deal of spluttering, on the hut, and rolled from thence to the ground; then another; and another again; and on turning round quietly I saw three jays on the grass, struggling to set themselves free; but the glued stick held them well, and the birds' fate was settled. In a moment more four or five more jays were also coming down, and Karl, withdrawing his fingers, allowed his 'call' bird to relapse into quietness. Thereupon those of the wild birds which were still free flew back towards the forest, settling on the border trees, and as soon as the coast was pretty clear of them, Karl rushed out, picked up the caught birds, and thrust them quickly into the large empty cage. There was a 'row' then, every one of the new-comers shouting most lustily, not only all the time they were held, but when caged; every time we as much as winked at them, they broke forth in the most unmusical and noisy of concerts.

"Of course, those that had gone away no sooner heard this shocking shindy, than they all flocked back to the rescue, and in less than a quarter of an hour's time over two dozen of them were also prisoners."

English bird dealers find that to take this bird no plan is more effectual than sham eggs as bait to a gin. They should be turned out of wood, birch answers very well, coloured and varnished to represent the natural ones. Thrushes are perhaps as good as any for the purpose, as they show well and are easy of imitation. Four or five of these eggs should be put in a sham or real nest, placed on a stage against a tree a few feet from the ground, leaving just room for the gin, which must have a little branch or two on either side of it, so as to bar access to the nest, save over the trap. The peculiar advantage of this plan is that, strange to say, it can be employed with success all through the winter when natural eggs are not attainable; and the false eggs can be carried in the pocket without fear of breaking them.

Then there is the magpie, of which old legends say, we read, that it still lies under Noah's curse, because when the other birds came of their own accord into the Ark, it alone gave trouble, and had to be caught. "What a delightful idea—the whole of Noah's Ark waiting to start, till Japhet caught the magpie!" It is everywhere a "fowl of mystery." On the far side of the North Sea it swarms, and perhaps does something towards keeping down the stock of game, for the "pyet" is desperately fond of eggs, and they often lead him into the gamekeeper's trap. On the shore of a shallow pond or lagoon which they frequent a small "pier" of stones and moss is built about three feet long from a shelving bank. At the end a steel rabbit trap is set. A hen's egg as bait having been emptied through a large hole on one side, a small piece of stick or a match with twine attached is placed cross-ways inside. To the other end of the twine a stone is fastened, and the egg is by this means anchored off the end of the artificial jetty. When a magpie sees the egg floating on the water, down it comes, and after a little while walks up the "landing-stage," to get within reach of the tempting morsel, and is caught in the act.

We have said nothing about the admirable chough, who, like the crow and raven, is faithful to one mate until death divides them. King Arthur's spirit went into a "russet-pated chough," the Cornish bards sing after Camlan; and the only mention we have in fable of the red-stockinged "market-jew-crow" is when he very properly refuses to "swop" his scarlet legs for the peacock's gaudy tail. He is modest and faithful in his personality, and attractive to the naturalist and lover of coast scenery. The jackdaws and hooded crows are as interesting as any of their kind, and fill up niches in the rich and varied bird life of the British Isles.

## CHAPTER IV.

# *MARSH BIRDS.*

### FOR POWDER AND PASTIES.

Plenty of sportsmen who are keen enough on heather or stubble "think small" of the various game of the marsh lands. They will when beating grouse cover pull trigger on a snipe, if he gets up out of a runnel, or springs from a miniature swamp where bog myrtles make diminutive forests on the wet peat tussocks. I have even seen these shooters "blaze" into a flock of plovers coming temptingly overhead, but it was always with an implied protest and a sense of the unworthiness of the game. Myself I do not much sympathize with them. Each sportsman will have his particular fancy in such matters, just as one man will go into ecstasies over a view which to another may be tame or barren. But, though not the first of sports, marsh shooting is very excellent in its way, and here at home, even in these days of hard draining and the reclaiming of land Nature never intended for cultivation, it is practised with enthusiasm and success by some of the keenest gunners on foot.

They may at least claim for their sport that it is universal and world wide. Other countries have their distinctive shootings to some extent. Pheasant shooters will not find much to do outside English shires, and he who loves the grouse must go to Scotland, while big game hunters look to Norway for reindeer, Canada for her moose and bison, India for the tiger, and Africa for the lion and elephant; but he who loves the merry brown snipe or his kindred, the wild wastes of sighing rushes, and the pools that flash back the daylight in their green setting, will find his delight in every land under the sun. When the ice cracks upon arctic tarns and the first fresh water of the year catches the red glint of the early summer sun, the snipe and the plovers are amongst those pools—very possibly there is no one to molest them, but that is not their fault; and as Lap or Siberian move up to the fells and spring time bursts over their country, all these little birds are there already. There is no better shooting anywhere, again, than in the tropics. The water-courses of the Flowery Land, the rich rice swamps of Ceylon, and the fertile plains of India have their Painted Snipe. A bewildering multitude of other game of much the same feather in Australia abound on the inland marshes.

Nearer home I am much tempted to dilate on many productive snipe-grounds, but the subject is too extensive. As a rule it may be pointed out rough shooting is best all over the Mediterranean and the lands bordering upon it, just at "flight time," and falls off rapidly after a week or two. But extraordinary sport is sometimes had amongst fen birds from Constantinople and the Black Sea to Cadiz, "while the fun lasts." Here, for instance, is a sketch of the sort of thing they sometimes get in the neighbourhood of Smyrna nearly opposite Cyprus; the writer being a resident and a good sportsman, his letter bearing a date in April. "A long and unusually cold spell," he writes, "brought our annual visitors in countless flights, to which the wholesale destruction apparently made no difference, as the cold intensified other flights, filled up the thinned ranks, until whole districts were alive with cock, every little bush holding a starved and emaciated tenant.

"In Smyrna the streets were inundated with thousands of cocks for sale at 3*d*. or 2*d*. each, or almost any price you would name. During one week it is computed twenty thousand were brought into this town; one French steam-packet alone shipped eight thousand cocks for Marseilles; in fact, the bird was a drug, and offering one of the most prized of birds to your friend was almost an insult. Many professional native sportsmen gave up shooting as not paying for powder, shot, and expenses, and on the first return of mild weather, when the flights returned to their breeding grounds, many might join in Dean Swift's grace, 'We've had enough.'

"Of individual bags, it is said one gun scored one hundred and sixty-eight cocks in one day; possible from the quantity and state of storm-driven birds, yet difficult to credit. Two guns, however, in fifteen and a half hours' shooting, extending over two days, bagged just one hundred couples. Other guns during a day, or even part of a day, bagged fifty and sixty cocks each—a feat accomplished by many ordinary shots. This took place in the neighbourhood of Smyrna, where the flights were more concentrated. If we consider that they extended over nearly the whole of Anatolia, and that thousands perished in the sea whilst crossing from the mainland to the islands of Chio and Metelin when storm-driven and feeble, the destruction must have been enormous, and may in no way revive hopes in your readers of future plenty. Yet over a wide expanse of country many thousands of birds never heard a gun; and if to these be added the apparent quantity of birds that survived the battering welcome of the elements and of sportsmen, which is known, it is clear that enough will return to the fens and lakes of Prussia, Finland, and Northern Russia, to breed numbers sufficient to partly make up for such destruction.

"A point for discussion suggests itself, however. Do birds that visit our shores, in case even of favourable wind and weather, ever migrate to England? Is it not rather those bred in Norway and Sweden which are welcomed by sportsmen at home? Flights from these latter countries will again vary the line of migration according to the direction of the wind at starting; so a scarcity in England may be occasioned from such a cause, and not actual decrease of the breed."

What joy would there be in Cornwall and Devonshire were such a flight as the above to grace their holly thickets and spruce plantations! But I fear we must wait for that fortunate breeze at the flitting season the writer suggests as regulating the cocks' movements. I have wandered rather far away, and possibly the woodcock of Smyrna, or the long-bills that teem in Caspian swamps are of little interest to home sportsmen. It cannot be denied that, as far as cock are concerned, there has been a lamentable falling off in the number of these birds to be obtained in our home shires—a decrease more marked indeed than that of snipe, though the improvement of land would on the face of it have been supposed to most affect the latter bird.

It is difficult to say why this is, though there are some causes which may be pointed to with certainty as having contributed to this undesirable end.

To begin with, the woodcock is a very shy bird, shy certainly in disposition if not in habitat. During the hours of moonlight and in open weather, like most of their kind, they are active and feeding in the open. Thus they have sometimes been taken in the poachers' drag nets when sweeping stubbles for partridges. You will find the woodcock during the day sitting under a clump of bushes or trees quite dry. If you examine the place where the bird gets up, you will see by the droppings it has kept its place after going to covert for the day, just as a hare keeps her seat until disturbed. They suffer a near approach the first time they rise, and should they escape being killed seldom afford another shot, unless a second person contrives to drive them to the gun, in which respect also they resemble hares, and it would not be difficult to imagine that the country where civilization is most oppressive to free spirits, and their midday repose is most often broken, would earn a bad name amongst them.

But more prominently than this cause of scarcity may be placed some others, the greater deadliness of modern arms of precision, the far greater number who use them, and lastly but not least, a consideration closely allied to the first one mentioned, greatly increased facilities of locomotion and consequent close search of spots to-day that formerly were natural sanctuaries unbroken by human intrusion. The general tendency of such altered circumstances would be to give England a bad savour at the woodcock head-quarters, to diminish migration, and would of course lead to the thinning of the ranks of such as ventured here and the practical extermination of those that might wish to remain and breed.

The woodcock, however, is a very persevering bird. With anything like fair play and quiet he will always find out favourite haunts and fill them. The Wild Birds Protection Act should have done as much for him as for any bird.

But with the mention of marsh or fen the bird that rises before us is the common and cheerful little snipe, which we have seen inhabits the globe from far north to far south. I doubt if, after allowing for the inevitable halo of romance which always tinges antique shooting stories, it could well be proved that our forefathers made much larger bags of snipe fifty years ago on localities which have remained unchanged than they could do in a favourable season of this or any neighbouring year.

"Capricious in their movements as snipe are, and influenced by every change in the weather, they are still fond of certain spots; and if they are to be met with in the country at all, you will be sure to find them in some of these. Snipe have wonderful 'lasting,' as an old gamekeeper used to say to me. 'Lord bless you, sir, I don't know how they stands; there's as many now' (this was just the end of the season) 'as if there was ne'er a one taken out of them the whole year—and sure we're shooting them every day since October came in.' It is a fact; shoot them as you will, there will be always some after you; and, though it is the fashion now to complain of the scarcity of snipe, I attribute that entirely to the great increase of drainage, as in localities which are favourable to them, snipe are as plentiful to-day as they were when I first began to shoot, now some thirteen or fourteen seasons ago. Of course, if you go in for improving the land, you improve the snipe out of it. Within a couple of miles of this there was a swamp with a river running through it, where in frost you could get your ten couple of snipe, with the chance of a duck or a mallard in the day. Now you would not see a snipe in it. An improving proprietor got hold of it, sunk the river, drained and sub-drained the swamp, and in two or three years had turnips and oats growing where I have often sunk below my hips shooting duck and snipe. Woodcock, I think, are getting scarcer every year, but not snipe; the only difference I see in the latter is that they seem to be wilder than formerly. This may be fancy on my part; I accounted for it by the mildness of the winters. This, however, I can say, that over the same ground I have got as good bags of snipe in recent seasons as I ever got in previous ones."

This exactly illustrates what I mean. "Of course, if you go in for improving the land, you improve the snipe out of it," but otherwise "over the same ground as good a bag of snipe is to be got now as in any previous season." I look upon the natural supply of snipe as practically inexhaustible, a result due to the infinite diffusion of the species and the vastness of their breeding ground. Here we shall never breed any adequate supply for the ever rising enthusiasm of the sportsmen of mud boots and retrievers, but from abroad we shall have a constant and unfailing fund limited only by the capacity of our counties to harbour and feed the strangers. "And both those capacities are getting narrower and narrower every day!" observes the pessimist.

There can be no doubt there is much truth in this. When King Alfred hid amongst the osier forests of Norfolk, and there was a current superstition that all dwellers near the Wash were web-footed, snipe must have found our littoral a very Elysium. To go no further back than the time of our grandfathers, there still live men who shot snipe amongst flags and reeds where there is nothing now but cobble and curb stones. Moorfields were an excellent place for long-bills, and so was Belgrave Square and Pimlico! There can be no doubt in the "home counties," at least, the available range of the snipe is within a measurable distance of total extinction, and it is so to a greater or lesser extent elsewhere.

Seeing, however, that snipe are to be had for the humouring, I would hint the advisability of pampering them wherever possible. At the risk of becoming an apostle of osiers, I would suggest, as in the chapter on ducks, the possibility of preparing with very little trouble tempting resting-places for migrating birds. An excellent letter to the *Field*, from one well and professionally qualified to express an opinion, illustrates the practicability of the idea.

"A Likely Place for a Snipe.

"Sir,

"While reading a pithy sketch of a 'white frost,' an idea came into my mind that has often arisen before, especially when crossing manors in different parts of England—how often the opportunity is lost of making a bit of covert for a snipe or a cock. I have met with lots of men belonging to my calling who knew well how to show a stock of hand-reared and indigenous game, that never once thought of establishing a bit or two in the place that would screen wandering game birds. It is simple enough; all that is to be done is to provide the 'feed' after having found the 'situation,' work a bit with the head, and then sign willing. I remember following a man on an estate in the east of England, and on asking him while we were going round the place, if they ever got many snipe, he said one 'now and agin.' Now, I had seen that it only wanted a little working to turn a stretch of about ten acres in a direct line into a first-class snipe beat, and all by commanding the water in the top pond by means of a sluice or two. This bottom ran as a fringe to the arable, and was between that and the coverts; and by judiciously breaking up, trenching, ditching, and planting willows, and on the highest places hollies, I could, in the flight time, generally about the middle of October (according to the season), keep my sedge and rushes in the top pond nearly root dry, and the water sufficiently low to afford food for the snipe, and by flushing the bottoms afforded future feed for them. The consequence was that, instead of a snipe or two 'now and agin,' I could all through the season show some, and frequently have seen killed by one gun down this beat, shooting over a small setter broken specially to work for snipe, a bag of eight or ten couples or more in an hour's easy walking in a white frost. The hollies would provide a cock or two, for if quiet they will not leave the feed far in a frost; and many a rare specimen this place has afforded—some of the most rare that have been obtained in this country—for in a few years I had a splendid reach, with about a dozen ponds of various kinds, some open, some well sheltered, and rough snipe ground between; for it is as easy to show snipe, cock, teal, widgeon, duck, and many of the divers on a shooting, as 'partridge, always partridge.' Many a sour, wet corner of a field, useless to grow anything but osiers, if broken up, trenched, and planted, would afford a snipe or two where one was not known before only as a *rara avis*. Find the feed, and my experience tells me they will go nearly anywhere. I could relate numerous instances of this. The great thing is to supply suitable and likely spots, and have them prepared when the first flight comes, and you can then stop those that would otherwise pass by you; and, feed once found, you may always insure birds. How is it a particular holly bush will provide you a cock as sure as the season comes round, but that the 'situation suits?' Can any one tell me why ducks and widgeon select different ponds, and keep to them, unless it is that situation has the main to do with it?

"W. J.,

"Noblethorp, Yorkshire." "Gamekeeper."

Ireland will always no doubt be a head-quarters for these dainty little birds, and they are still to be had in more or less abundance every spring and autumn along our eastern coasts. There, too, are reed beds and gorse dunes that hide in season other quarry for the sportsman naturalist—happy hunting grounds once of the fen netsman. Daniel describes how when a fowler discovers a marsh hillock where the ruffs and reeves play, he places his net overnight, of the same kind as those called "clop" or day-nets only, generally single, and about fourteen yards long by four broad. At daybreak he resorts to his stand at the distance of one or two hundred yards from the nets—the later the season the shyer the birds, and he must keep the further off. He then makes his pull, taking such birds as are within reach. After that, he places stuffed birds or "stales" to entice those that are continually traversing the fen. "A fowler has been known thus to catch forty-four birds at the first haul, and the whole taken in the morning was six dozen; though, when the stales are set, seldom more than two or three are taken at the same time."

Mr. Lubbock, in his "Fauna of Norfolk," says that in that county nets were never used to take this bird, but rather snares made of horsehair. Then again there is "the foolish dotterel." In the whole range of English poetry, only two writers mention him; one is eccentrically unfortunate in his remarks, and the other draws heavily upon the recognized licence of his order. Wordsworth in the "Idle Shepherd Boy" writes: "the sand lark chants a joyous song." Now that appellation is a local name for *Charádrius morinellus*, and we need not say that the shepherd boy would deserve any other title but that of idle who caught the silent dotterel chanting any sort of ditty. Drayton again, in "Polyolbion," tells us—

"The dotterel which we think a very dainty dish,
Whose taking makes more sport as man no more can wish,
For as you creepe, or coure, or lye, or stoupe, or goe,
So marking you with care, the apish bird doth soe,

> And acting everything doth never mark the net,
> Till he be in the snare which men for him have set."

The poet has here clearly accepted as fact exaggerated stories of fen men, based no doubt though they have been on substantial facts; for this "shore lark," when newly arrived from the primitive barbarisms of the far north, is both foolish and curious. Bright lights in the darkness of night possess an attraction for birds often taken advantage of for their destruction. Thus the dotterel was caught in long fine nets extended about the marshy sheep-walks frequented by them, of mesh just sufficient to admit their heads, and supported on light sticks. Then when the birds were settled for the night, and land and sea were merged in equal darkness, only divided it might be by the lines of pale breakers running in upon the shores, the dotterel men turned out, some walking in line beating with their sticks, and others clanking round stones together, and the drowsy birds ran before them. Another party stood just behind the nets with lanterns, and attracted by the glare of these, the birds ran towards them, and were speedily entangled.

The knot again, that bird that was to King Canute what lampreys were to John, is found amongst the sedges, and has been decoyed into nets by wooden figures, painted to represent itself, placed within them, much in the same way that the ruff was taken, the best season for their capture being August to November. We doubt if a dozen a year find their way to Leadenhall Market now, so much have they gone down in fashion or in abundance.

But I must not run through the whole gamut of unrecognized game which lives outside the manor and beyond the pale of an ordinary shooter's sympathy. There is the heron, that spectral blue bird of preternatural sagacity, and the bittern—not yet quite extinct—whose weird cry doubles the loneliness of the swamps and wastes. There are coots and moorhens which are wise enough to be equally unpalatable and sombre feathered; the grebes, quaint in manner and form, the dainty godwits who dabble in the brackish pools or flit lightly down the nullahs, not to mention the curlew and his kind, or the plovers who wheel and scream in the yellow of early dawn overhead. But these birds are for the most part unpopular, so they get but short shift.

I think, however, with the man who loves snipe and sedges, that there is good and healthful sport to be had

> ". . . by the drear banks of Uffins
> Where the flights of marsh fowl play;"

and in union with him, as well as from early association, the wild birds of the river-side will always be appreciated by me.

## Winter on the Mud Flats.

To make a successful marsh shooter, capable of enjoying the lonely wastes, even though we indulge in this fascinating sport with the best regard to health and comfort, demands good health and a certain amount of hardiness, tempered by judicious caution. The way in which the most pleasure can be obtained, with the minimum of discomfort, is unquestionably by shooting from a boat, especially in the season of frost and snow, when it is no mean consideration to have a dry shelter to retreat to always at hand. The boat-shooter thus may find himself quartered over night at some comfortable water-side inn near a favourite haunt of the cold weather birds.

His next discovery is that "five-o'clock-in-the-morning courage" is one of the essentials in the character of a successful rough shooter when he indulges in this frigid pleasure. His slumbers between the sheets of his comfortable crib hang entirely on the state of the tide; perchance he is just indulging in that "beauty-sleep" which doctors tell us it is such ruin to break, when there comes the rattle of gravel on the lattice windows, thrown from the hands of a grim old "salt" below, who appears to sleep in his thick blue jersey and rough cloth breeches from year's end to year's end, since it is impossible to notice the smallest difference in their arrangement no matter what the weather, or how unearthly the hour at which he plays chamber-maid. The early morning when the shooter reaches the snug coffee-room—where the "neat-handed Phyllis" has already lit a roaring fire with the ribs of some long-ago wrecked vessel, and made preparations for breakfast—is remarkably cold, the windows are dimmed with frost along their lower margins, and everything outside is silent, chill, and grey. As far as the eye can see down the little village street which ends in "the hard," and a muddy creek, whence fishing-boats gain the open water of the tideway, no soul is stirring—the very boats are asleep, waiting for the water and the tardy light to open in the east. Breakfast over, the old sailor is followed down to the water's edge, where he deposits his burden of guns, bags, and wraps under the deck of a little craft that lies on her white-painted side just awash of the tide. She is soon shoved afloat, and, with a rag of a sail, goes creeping down the creek, her skipper at the helm, and the gunner forward, seeing all clear amongst the stowage and lumber ere "going into action."

It is still very early, and the sportsman feels much inward pride at being afoot so long before the world is awake or many folk have shaken off their slumbers. He may be conscious of a zero temperature about toes and fingers, but he does not care for that. There is a prospect of really good sport before him, for the mud flats are just being uncovered by the still falling water, which leaves in its rear wide stretches of ooze, rich in soft-shelled straddling crabs, incautious flat-fish, and other marine delicacies, the presence of which is tempting the sea-game down from the marshes, where scores of them have been bickering and whistling during the evening.

If the shooter is of still hardier mould he will have slept amongst these "noises of the darkness," and in spite of feather bedsmen, there are worse places on a frosty night than the cosy cabin of a fishing smack. It is true accommodation is limited, and the landsman will look round helplessly for the conventional hat-stand, besides being likely enough to suffer from low hatchways, and to feel generally "cabined, cribbed, confined" for a time until he has got more used to the limited space 'tween decks. But for those to whom winter shooting is the best in the year, who love the tonic sting of a north-easter as it comes blustering over the salt flats, hurrying down a whole new fauna of bird life from the breeding grounds of the far north, such hardships deserve a gentler name when leavened by prospects of a few hours' brilliant sport on the morrow.

A frost of a bitter kind came on not a dozen winters ago, our coasts being peopled for a week or two with wild birds, of a score of species, in flocks the like of which had rarely been seen before. The cold, while it lasted, was Siberian. We had chartered a handy fishing vessel, used once or twice before on such occasions, to await us as near as she could come to an out-of-the-way station on a sea arm that ran in from our north-western coast; and by nightfall on the third day of the frost we were rid of most of the conveniences of town life, and afloat 'tween decks on our smack. A warm at the gallery stove, a pipe, and a pannikin of the skipper's after-supper "tea," a yarn or two more or less spiced with the improbable, always nourished by sea air, and then a few hours of sleep under the yellow glow of a swinging lantern, were the preliminaries for next day's work. An old hand under these circumstances, coiled in his sea-jacket, a good blue jersey rolled up for a pillow under his head, and comfortably swathed in a stout Witney blanket, will sleep the sleep of the just, in scorn of down beds and the frost outside. If he has served an apprenticeship to green waters he will know in his slumber when the tide has turned as surely as though he had sat up to watch, probably going on deck to have a look round. At two a.m. we thus turned out to see what the weather was like. There was an arctic stillness in the air and almost an arctic dryness. The wide sweep of sky overhead, dotted by a thousand stars, was glittering with wonderful brilliancy, and the gleam of the land under its snowy mantle just showed its whereabouts. The only sounds audible were the noise of thin ice floes grinding together in the filling creeks, the tinkle of salt water falling through sluice gates, and, tuneful to the listening wild-fowler's ear, came the sound of the moving birds feeding and flying over salt wastes and estuaries, the chorus of the ducks inveighing against such "hard times" near some water hole, the whistle of restless widgeon, varied every now and then by those wildest of all sea bird's notes, the "troomp" of wild geese, or the unearthly booming of a heron.

By six o'clock we had thrown off our frozen warps and dropped down the creek as the day came. We stole along quietly under the high banks, making hardly a sound in the still cool air of the morning, until over our sorrel, samphire, and sea grass sky line, the skipper sees a couple of curlew coming up, and putting us on the look-out by an expressive wave of his short black pipe, tight between his lips since "the ship" got under weigh. Both curlews pay the penalty of their rashness, coming down headlong into the water with a loud splash, and are brought into the boat as she goes by with the help of a landing-net on a long staff, a fair beginning, since the old rhyme says that

> "A curlew, be she white or black,
> Still carries tenpence on her back."

Then a shingle point is turned, the boat sliding into more open water, where she "goes about" and steals up as near as she dares to the edge of the flats. The skipper's practised eyes soon make out a cluster of black dots half a mile to leeward under the veil of mist which still hangs over the river, and a look through the glass shows them to be ducks beyond doubt. Not a shade of any emotion exhibits itself on the honest face of the boatman as he cautiously edges his vessel down to the knot of birds that are feeding amongst the masses of floating weeds, taking care, however, to keep her a point or so off them until, when the suspense is at the highest, the helmsman broadens her off a little, and as the birds, uneasy and at last affrighted, crowd too late together before taking wing, big gun comes to bear straight on the "brown of them," and the old sailor, with a nod of approval, fills the sails again, bringing the slain within reach of the landing-net. Those only winged require some skilful manœuvring and a cartridge apiece, before they are laid beside the others under the thwarts.

This is a very satisfactory continuation of business, but there is still plenty more work on hand, so pipes are hastily filled, while the tide and a slant of wind drifts the boat up the estuary to search for another flight of birds. By this time the sun will be up, lighting with a vivid glow the red sprit-sails of a convoy of barges, and dispelling the thin drapery of vapour that has hitherto hidden the opposite shore, which now, however, starts up into light and shadow; while the water takes a new tinge from a sky of roseate pearl overhead.

A wisp of ox-birds got up, almost under the button at the end of our little bowsprit, going twittering down the water as though they would never stop, but we reserved our powder for better game. This was not long in coming. In passing the mouth of one of the tortuous watercourses, draining down into our main channel, a couple of teal flew within easy shot, one of them being stopped promptly, and then the other. Both were speedily brought to hand by the retriever we had with us, who seemed greatly to enjoy his plunge overboard and the scamper over the dead weed and samphire flats. At the shots, a heron rose majestically, and then passed swiftly out of our range. A company of widgeon also went away to open water, and a cloud of golden plover rose on the wing, flashing in grey and white as they wheeled hither and thither; but what absorbed the attention of our skipper most, and riveted his keen sight for a minute or two, was a glimpse of, perhaps, some dozen birds of more than ordinary size, that took wing, and after a turn, settled again heavily in a creek about a mile away. It did not need his brief ejaculation to tells us they were geese, and forthwith, all lesser game was forgotten in the prospects of a shot at these choice birds from the far north. We stood across the shallow salt lagoon, just as the red winter sun was coming up in the east, the fresh north wind coming with it until we were half a mile from the solitary gander standing sentinel over his flock on a mud bank, outwardly engaged in sorting his back feathers, but, as we knew, keeping a sharp watch on us. Then we got the little dingy to the yawl's "off" side, and ourselves, boy, and dog, slipped in unobtrusively, but held on to the ship until the skipper edged us under cover of a mud bank. We pulled ashore, and while the boy minded the boat, we ourselves slipped in a couple of No. 4 cartridges, previous to making a careful stalk of some three hundred yards. We were within forty paces of the sentinel, on our hands and knees, when his contented chuckle gave way to the silence of alarm. In another minute the whole flock were off in their cumbrous fashion—all but two that we had stopped, one of which gave the dog a lengthy chase.

These mud flats are dangerous places to the careless. The waters fill the gullies so insidiously, that one may well be cut off and drowned, unless the greatest precaution is taken. With boats in attendance it is another matter, of course; but it was at a lonely landspit, covered by a fathom or so of water at high tide, on these shooting grounds, that the body of a shore shooter was found only a short time ago. The luckless fellow had clearly been separated from the mainland, and had gone to the highest ooze he could see, had driven the barrels of his gun into the mud, and tied a leg to this feeble stake to save himself, it was surmised, from being washed away, meaning, probably, to stand the flood out; but the exposure and cold proved too severe. He had been dead several hours when a lobster boat found him.

But time and tide will not allow much space just at this juncture for melancholy reflections, and the skipper draws the attention of the "gents" to a couple of widgeon that are coming up wind at a great pace. It is doubtful whether they will pass within shot, but fate has marked them, and they sweep nearer and nearer. Some fifty yards away, one is bagged in very good style, though the other gets off scatheless, untouched by a hasty shot. Then a heron goes over the water to some fishy pool he wots of on the far side, with neck folded back and long legs trailing behind, as high above the world as an aeronaut making meteorological observations. Sandpipers succeed singly or in flocks, and subscribe a victim or two to the bag; lapwings, tame and silly, also paying dearly for their disregard of ordinary caution, and their cousins, the golden plovers, more business-like, wheeling hither and thither on rapid wings, showing their numbers clearly, or becoming almost invisible, as the position of their bodies varies against the dark background of the saltings. These, and many other birds that winter sends to gratify the rough shooter, people the estuary and afford shots more or less exciting, from sunrise to sunset.

While the puntsman fires his big gun, it may be but once in twenty-four hours, the wild-fowler, who uses a breech-loader, has better and more exciting sport—or at least more varied. A strict chronicle of his day's work in a good river, in a hard frost, would be a difficult task to undertake. Each shot he fires is distinct in itself, and the pleasure of working up to his birds, and the knowledge he gains of the curious ways, are often keener than the final successful result of his shot.

A taste once conceived for such sport, as that I have

attempted to outline, is very difficult to eradicate. It holds its votaries from youth and rashness, to age and rheumatism; it is never possible to have too much of it, and enthusiasts declare every day's sport is totally unlike the last, wherein, perhaps, lies some of its charm.

Truly it is pleasant again in June to lie far out from the world amongst the long grass of some shingle pit—

> "With the winds of summer blowing,
> O'er the wide sands wild and free;"

and for once not bent on destruction, but in pleasant fellowship with lowly nature to watch the wild bird life—those dainty redshanks, for instance, who glitter and flash in the sunlight as it catches their white under plumage before they settle in a piping cluster on the flats. They dabble with infinite daintiness their coral beaks and legs, keeping them spotless and immaculate in a world of sludge. There are many other birds as pleasing and curious in their ways, of which no one but the cockle men and the marsh gunner know really anything.

The snipe shooter proper is a being of higher sphere; he rarely mixes with the Bohemians of the banks. Yet his special pleasure is, as we have said before, very delightful and engrossing. His is a fine art in itself—an art that can only be learnt in its fulness by long years of patient study. Where the snipe will lie to-morrow, and where they will come from, whether to beat them up wind or down wind, with dogs or without, to take them in the rain or, with the gallant Colonel Hawker, to wait until the wind has dried the rushes; are all important and pregnant questions. The themes under the heading of "snipe," are infinite. Who shall say with any finality whether No. 6 or No. 8 shot is best for him, whether it is true he uses his bill in rising, whether he really listens for the creeping worm, as he certainly seems to do when that delicate head of his is turned on one side,—even of what his food consists, and whether August is too early to begin shooting.

We can only be glad such an admirable little game bird is still left to us, and if our brothers of smock frocks are each to have "three acres and a cow," then we respectfully petition in the name of those who love the gun, that our portion may be, forty acres—and a snipe!

## Why Snipe are Scarce.

Though the misguided English yokel, who is to have the heifer and the triple meads as his share, may do much damage to our native wildfowl haunts, I doubt if he is responsible for the death of all those birds unmarked by shot which wave in the wind over our purveyors' stalls like the companions of Ulysses in the Cave of the Cyclops. It is rather the ingenious and mercenary foreigner who sweeps his fens and hill-sides to cater for the discriminating taste of "mi lord Anglais," and sends us "poached" snipe and woodcock by the crate full. Even our good cousins across the Atlantic, now ice rooms or refrigerators are fitted to nearly all steamships, evade their none too stringent game laws, dispatching us netted wild birds from Chesapeake Bay and the wonderful rice swamps of the interior. More than a few of our Leadenhall wild geese and ducks have come from the Yankee shores, and even, perhaps, that turkey who makes a final appearance at our Christmas boards may hail from the chippy curtilages of Canadian squatters' wigwams or the adjacent snow-buried pine forest.

It is clear, for instance, when we read in weekly market reports how woodcock are selling at a few shillings a brace, while under the same date a sporting paper goes into ecstasies over the fact that a single couple of these little winter visitors have been flushed from a south coast spinney, that the market must perforce be supplied from some other source, and we should look abroad for it. Not so very long ago Cornwall and Devon were equal to the epicures' demand, and Exeter coaches of the day used to bring as many as thirty dozen in a week to London. One person, an old writer tells us, sent in a single season from Torrington, in Devonshire, woodcock to the value of £1900 pounds into market. Truly those were the days when "cock" were at the height of fashion, and ten, sixteen, and even twenty shillings a couple was willingly given for this admirable table bird. At that time woodcock were taken in the south of England by V-shaped enclosures in coppices and woods they frequented, formed of small light fences of dead holly or beach boughs a foot or so high. The woodcock, instead of attempting to leap or fly over these, ran down the inner side, looking for a small opening to creep through. This he found at the apex of the angle, but a noose hung over it which effectually secured him by the neck—a victim to undue fastidiousness!

But that Torrington game-dealer would never have made an annual income of four figures out of *Scolopax Rusticola*, had he known of no other snare but the above somewhat "single-barrelled" affair. The glade-net was no doubt the engine with which the western men took most of their quarry, though the device, I am well pleased to think, is hardly ever used now on this side of the Channel. It consisted of nets hung across the open rides in coppices, "the cock roads," as Blome calls them, into which migrating woodcock, and sometimes partridges, and even hares, plunged when driven from the neighbouring woods by beaters. "The nets have to be of length and breadth proportionate to the glades in which they are suspended," says Folkard, in his "Wild Fowler," a volume that should be on every sportsman's bookcase. The net is suspended between two trees directly in the track of the woodcock's flight. Both the upper and lower corners have a rope attached to them, which is rove through sheaves fastened to the trees on either side, at a moderate height, varying from ten to twelve feet. The falls of the two upper ropes are joined, so that they form a bridle, to the central part of which a rope is attached held by the fowler in his hand in a place of concealment, and thus he is able to drop it down suddenly and intercept any rash bird hurrying down the drive, the working of the net being assisted by five pound stones tied to the corners. The fowler having stationed himself in such a position as to command a full view of the glade, beaters are employed to flush the cocks out of their retreats amongst the dead fern and undergrowths, if they are not actually migrating at the time, and just as the bird approaches the net it is suddenly let down or drawn out. The instant the birds have struck the net, the fowler lets go another cord looped to a stake within reach of his arm, and the whole net, with the birds entangled, then drops to the ground. In France they are particularly skilful in this art of taking the "bécasses," and the glade nets they term "la pantière."

All birds when migrating fly through gaps in mountain ranges or any alleys natural or artificial assisting their progress. It is as if, conscious of a long journey before them, they took advantage of every chance to avoid digressions or deviations from the straight line. Thus, in sweeping over Heligoland, woodcock often pass actually through the streets of the town, and the worthy burghers, taking advantage of this, hang out from window to window at nightfall fine twine nets of small mesh. In these next morning, if the towns folk are in luck, hang an assortment of cock with, perhaps, a sandpiper or two! There are even stories current that men out after dark have been knocked down and half killed by some blundering mallard or errant pochard taking a short cut down the local high-street in its autumn flight; but the narrative wants confirmation. "Now is the woodcock near the gin," says Fabian, when Malvolio stoops to pick up the forged letter of his mistress; no doubt the bird has been harried and hunted in one way or another from time immemorial. Perhaps there is no method more eccentric of taking this foolish bird than that French fashion, "*à la folatrerie*," we read of in "Le Moyen Age et la Renaissance." The fowler had a dress of the colour of dead leaves; his face covered with a mask of the same hue, having two holes in the place of eyes. As soon as he saw the woodcock he went upon his knees, resting his arms upon two sticks to keep himself perfectly motionless. Whilst the woodcock did not perceive him, he advanced gently upon his knees to get near the bird. He had in his hand two small *baquettes*, the ends of which were dressed with red cloth. When the cock was stationary, he gently knocked the *baquettes* one against the other; this noise amused or distracted the attention of the bird; the fowler approached nearer, and ended by casting over its neck a noose which he had at the end of the stick. "And know this," adds the French writer, "that woodcocks are the most silly birds in the world." No doubt the foregoing discredits their sagacity sorely, but quails in Afghanistan, as many travellers point out, are caught in much the same way. There a native sportsman dons a yellow shawl with large black spots, and by crawling on "all fours" into the barley fields or peach orchards, palms himself off on the credulous and curious birds as their mortal foe, a leopard, whom they surround and mob. At first sight none of the three species of our lesser snipe, the common bird of rushy patches, the gamey little jack snipe, or the scarcer great snipe, would seem to tempt the fowler's art. Erratic in habits, and curious in feeding grounds, there is no knowing with any certainty where to look for them at a given period. Watercourses and the little "canôns" draining moisture from marshes or meadows, are likely spots in frosty weather. There the country people in Ireland catch a good many snipe in what they call "cribs," which are a kind of basket, roughly made of pieces of stick tied together in the shape of a pyramid. This is supported by an arrangement of forked sticks very similar to that used for the old-fashioned brick trap. This crib is set by the side of a spring, and a snipe going inside it releases the catch, and the basket falls over him. The country people say that if a snipe is left any time in a crib, however fat when caught, he will get quite thin from fright and his attempts to escape.

This latter fact is curious, but quite credible, seeing how under opposite circumstances this species of bird plumps up with even a few hours good feeding after his autumn migration or a period of frost starvation.

Amongst reeds and rushes are sometimes to be found little paths pattered smooth by moorhens and water rats running to and fro amongst the stems. Here, Sir Ralph Payne Gallwey tells us, a springe for taking snipe, woodcock, and other wildfowl, often used in Ireland, is made thus: Stick a pliant wand of a yard and a half firmly into the ground, bend it down till the ends of a short cross piece attached to it, and which may be four inches long, catch in the notches cut to receive them in two stout pegs driven firmly into the ground, and showing a couple of inches above the surface. Pass the fine wires that are attached to the cross stick over a slight nick in the top of each peg, and place the running nooses flat on the soil for snipes, edgeways for ducks and teal. When a bird is snared, the little stick crossways between the uprights is freed at once, the wand flies, and the victim is strangled. This is done so quickly and quietly that the captive is not missed by his companions, though dangling above them. He has found half a dozen duck, teal, snipe, etc., thus strung up in a morning!

Such are some of the illegitimate devices tending to make both snipe and their big relative, the woodcock, scarcer year by year. No doubt there are more wholesale methods such as the fen men's long nets which go over some fens, especially near large towns, in the dusk of the evening, and frighten away what snipe they do not secure. Possibly not quite so much is done towards making these little wildfowl at home in our waste lands as might be. There are, as said, scores of places on many estates, even in the midlands, which by a little preparation in the way of flooding a corner or two of "debatable land," and putting in a few willow bushes for cover, might be made to hold twice as many couple of snipe as they do at present. I take it as a fact not to be denied, there are always plenty of wild birds somewhere to occupy a desirable spot directly it is formed; but the worst of it is, desirable locations are becoming so sadly scarce in our over-drained and over-"improved" shires!

It is in any case certain, however, we cannot be wrong in suppressing nets and kindred engines of wholesale destruction, and giving the snipe a chance of resting in the limited selection of osier beds and marshes yet open to him.

## CHAPTER V.

# GROUSE.

MOOR AND MOUNTAIN FOWL IN THE THREE KINGDOMS.

All the grouse family interest the sportsman more than the agriculturist. Hill crofters who live amongst the grouse bear no great grudge against them, though now and again perhaps a brood will come down to search their rocky stubbles for shed corn, even pillaging to some small extent the barley shocks after harvest. But the damage they do is trifling at worst. Foresters complain that capercailzie and blackcock eat the tender shoots of silver firs in "hard times," but the complaint is not of any great weight. Radical legislators, perhaps, who believe in breaking up estates to instal a yeomanry on allotments, and would like to see the gneiss of Ben Nevis planted with turnips, and the sides of the Grampians devoted to carrot plots, may bear ill-will to the whole race; but their arguments are more trivial than either of the others! To gunsmen grouse stand at the head of all our indigenous birds. They are to them what the salmon is to fishermen, and the elephant and tiger to Indian sportsmen. How welcome is the eve of the 12th of August to those whose fortune it is to have toiled through a long hot summer amongst city dust for the rest of northern moors! That night journey itself that takes us northward, with all the luxury of modern travel, is, the first time we make it, an experience the fascination of which never fades. There is the wonderful rush through the fertile midlands, the chequered landscape under the moonlight, the long gleam of lights of sleeping towns whose names we can only guess at as we fly over the faultless steel roadway, and then the lurid flare of the furnaces down the vale of Trent. We have had our hot coffee, and taken our cigarettes, and perhaps "forty winks" in the folds of our thick ulsters, when dawn comes in the east over the deep dells and stone walls of Cumberland; and classic Ridblesdale, that most fascinating valley, holds us before the sun has melted a single dewdrop or thawed the thin white frost that silvers the shadows. What does it matter that we have lost a night's rest, and that we are perchance somewhat travel-stained? The Border is at hand, and beyond it lie heather lands and those grouse we have thought and dreamt of during weary days at work and the dusty crabbed hours of endless sessions. To-night we shall be amongst the hills, and to-morrow we shall breathe again air that is worth inspiring, and look upon scenes that are a tonic and a sedative—a very lethe of happiness to a hack of dusty civilization.

In fact, grouse shooting has a special charm of its own. It can never cease to be popular in one sense of the word; while regarding its accessibility—to any but the wealthy—it must be confessed that the recreation and all delights it brings with it seem to be going back into the regions of the impossible.

There was a time, and not so very long ago, when a tract of moorland reasonably stocked with its own natural grown birds could be had, if not actually for the asking, for about the price of a week's shooting in the southern shires to-day. There was more adventure and more sport, I think, under those circumstances, and decidedly more of roughing it in the style of which Scrope speaks so enthusiastically. It was a tour of many changes, from mail-coach to dog-cart, and trap to pony, at those times, to reach outlying moors; the shooter and his friends provisioning themselves for a siege, moreover, like Border raiders when the Wardens of the Marches were out. Plenty of exercise was then a certainty, and the honest old muzzle-loader being in each man's hand, the grouse had a better chance of making good a retreat; men shot in more instances with greater moderation, more pleasure in the shooting, and less in seeing the total of their bags in the next batch of local papers!

In those Arcadian days, when the shooter was not deposited in the midst of his grouse land by luxurious sleeping-cars, which had brought him north from the metropolis in nine hours or so, such a thing even as free shootings were not unknown. Amongst "the islands" and the rocky glens of the western coast, a man might establish himself and roam pretty nearly at will; to-day I doubt if there is a grouse in the highlands that could be shot without leave by any man of conscience, for ownership has extended in every direction, and the happy debatable lands of our forefathers are known no longer.

I have no desire whatever to decry grouse or grouse preserving. The teaching of our demagogues that moor and mountain belong to the peasant and should be cultivated for and by him alone is difficult to refute, because there is a grain of truth in it. Some seventy per cent. of the highlands cannot and will never be cultivated by any crop that the crofter can afford to rear. Such soil, rock it were almost better to call it, is fit only for grouse and the slow-growing firs and spruces (harbouring capercailzie and black-cock) which give no return for capital for twenty years. As for the remaining percentage of land, much of it is cultivated. If it *will* grow crops and does not, then it ought to. It is on this peg of a little cultivatable land uncultivated that agitators hang all their grievances; and land-owners would do wisely by taking the ground from under their feet and helping crofters to reclaim that strip of bog they covet, and to build a cot to look after their poor harvest of ragged grain. The shame of the highlands to-day, and their pressing danger during the next ten years, are the few incorrigible landlords whose views have not broadened with the times, and who would tyrannize in mediæval style over a long-headed and thoughtful yeomanry who are germinating new ambitions under the light of better education.

It is such, and the harshnesses of American millionaires, who oust pet lambs from cottagers' paddocks, and depopulate glens to keep a few more head of deer, that endanger our northern shooting and strengthen the hands of demagogues. If the Game Laws are ever abolished, and we lapse into the gameless condition of France, for instance, it will be the direct result of such game-preserving as this.

As for those heresies of a higher class, the erection of "trap" fences round deer forests, by which your neighbour's stags can join those which are legitimately your own but can never return to their own feeding grounds again, and the snaring of your brother sportsman's grouse by nets put up along his marches, they are offences of the deepest hue, bar sinisters on the sportsman's escutcheon which should place him beyond the pale of any friendly intercourse or good fellowship, and reduce him at once to the rank of a professional poultryman.

Such measures as the Access to Mountains Bill and others affecting game preserving will come, and ought to come shortly; but otherwise, I think there will be no very revolutionary game legislation for a long time, no matter how much Radicals may bluster.

As to the *natural* prospects of grouse, as one shooter observes, it has become almost a custom of late years on very prolific moors to test the number of grouse killed by comparison with the figures of the year 1872, the greatest grouse year ever known both in Scotland and in Yorkshire. In 1872, on a famous Aberdeenshire moor, 412 brace were killed over dogs, by four guns, on the 12th; within a fortnight of this unexampled performance, Lord Walsingham killed his famous bag of 423 brace to his own gun, in one day, on his moor of Blubberhouse, in the Otley district of

Yorkshire; 1000 brace were killed in one day at Studley Royal, 1100 brace at Wemmergill, and, to crown all, the highest yet recorded bag of grouse in one day, viz. 1313 brace, was made at Mr. Rimington Wilson's moor of Bromhead in the Sheffield district.

During the same fortnight, 10,454 grouse were killed in ten days' shooting at High Force; while at Bolton Abbey, for eighteen consecutive days, the average of grouse killed per diem was within a fraction of 300 brace. These figures sound like romance; yet their writer vouches for the accuracy of them all, and they are well known to be correct by those versed in the figures of northern moors.

The Scotch moors in the same years yielded phenomenal results.

By way of comparison, let us take an instance or two from the Yorkshire records of this year. On Danby moors, belonging to Lord Downe, after killing 600 brace over dogs, 900 brace were killed in three days' driving. On Wemmergill, Sir Frederick Milbank and party, six guns in all, slew in six days 4523 grouse, or an average of 376 brace a day. At Bromhead, in the early part of September, over 600 brace were killed in one day.

These figures, it will be noted, are but for a few estates. The produce in grouse of even a single Scottish shire is infinitely greater, and represents a very considerable amount of human food. "But," says the illogical stump orator, "it is food only consumed by one class." To this it should be said that, philosophically, the class who can afford to eat game, by doing so sets free other less expensive food for another section of the public.

But it is the influx of visitors, the trade, and the briskness they bring with them that must be chiefly held to benefit the highlands, and socially justify the devoting of wide wastes to the muir-fowl. Let it be always remembered, and the evidence is at hand in Government Reports, that the farmers, large and small, of the great English game-rearing shires, turned out to be the warmest supporters of game-rearing and preserving when examined before the Game Laws Committee of the House of Parliament. It is the orators of Manchester allies and politicians of the salubrious slums of Chelsea who object on principle to property in fur or feather.

The rents of shootings are far too high. This is, of course, the result of keen competition. Worse still, the competition extends itself into the actual shooting, and when an agent stretches a point and says, such and such an estate ought to produce five hundred brace of grouse, the owner for the time naturally likes to get his thousand birds, and grumbles if he doesn't. A more fatherly interest is what we want for our northern shootings, and less, far less driving at the very end of the season. Mr. Archibald Stuart-Wortley very justly remarks it is not only outside warm corners of Suffolk coverts that "bird butcheries" take place, they are known sometimes on the far side of the Tweed when hot autumn days make the grouse lie in the bents like quail under a hedge, and the breechloaders mow them down at half distance remorselessly. I do not agree with Mr. Stuart-Wortley in his opinion that Scotch moors can never again carry such a head of game as they have done; though agreeing with him that during the last five years they have been shot, "not wisely, but too well"—with too much science, and too skilfully—for "the pot," or worse still in some instances, "for the poulterer!"

Everywhere firs and spruces are being planted along the straths, and this should tend to the increase of that noble bird the capercailzie, who is rapidly regaining his position amongst the lochs and corries. A like cause should tend to the multiplication of blackgame, who love the openings in these plantations and the hollows overgrown with cotton grass and willow. As recorded by Mr. E. Harting and others in the *Field*, many attempts have been made to introduce the blackcock into Ireland by the importation of living birds of various ages, but entirely without success. The hatching the eggs under pheasants appears to offer a very reasonable hope of permanent success. The young would be accustomed to their new locality and the peculiar food furnished by it. I do not believe in the impossibility of introducing black game into Ireland. For a species whose habitat extends from Scotland in the north to the New Forest on the south coast of England, there must be found many suitable situations in Ireland; and the varieties of food on which it feeds are equally abundant in both countries.

Landed proprietors should try to introduce this noble game bird into different localities in the Emerald Isle, selecting, as offering the greatest hope of success, situations similar to those affected by the species in Great Britain—not barren heath and moorland, but the vicinity of woods, coppices, and semi-cultivated lands, where alder, birch, and willow twigs are found in the spring; crowberries, whortleberries, and similar fruits in the autumn; heath and vaccinia all the year round.

The ptarmigan, that bird that conspires with the seasons to hide him, is, I greatly hope, able to take care of himself amongst his rocky fastnesses. None of these birds do any recognizable harm to the produce of human industry. As for the red grouse, I see no reason why he should not flourish and multiply in face of the grouse disease and human and feathered foes.

Those who are fortunate enough to be able to spend long days in August or September upon the heather must, however, be moderate and philosophical in their sport, or we shall be within measurable distance of exterminating one of the finest game birds in the world and ruining a valuable recreation. Wise protection is also essential, and the stern suppression of unseasonable poaching.

A matter that ought, for this reason, to be of some curiosity to the sportsman, is the abundance of capercailzie, blackgame, and grouse hanging in rows along the outside of our poulterers shops late into every spring, and attracting attention by their cheapness. It goes almost without saying, they cannot all be English birds. Even their purveyors would hardly pretend that. Whence do they come? The answer is, from abroad; the capercailzie and ptarmigan from Norway and Sweden, the blackgame largely from Russia, but the red grouse undoubtedly from the northern part of our own kingdom, as *Tetras Scoticus* is unknown elsewhere—the only creature in the British fauna that can lay claim to that exclusiveness. Winter is a bad time for all these birds, and the snow-covered ground which sharpens their hunger and brings them into the snares of the poacher, also lets the gunner in—a worse poacher as often as not than he of the nooses and nets—by betraying their hiding-places and showing up their crouching forms.

In autumn the capercailzie of a district are divided into packs of fifty or a hundred, the hen birds keeping separate from these gatherings, which feed along the sides of the numerous lakes and morasses with which the northern forests abound. The Swedes shoot these noble birds by torchlight.

During the winter many capercailzie are also taken in snares. Two stout sticks, some eighteen inches in length, and forked at the upper end, are driven into the ground on either side of a pathway, and across these a third stick is placed, from which depend as many nooses of horse hair as may be convenient. The nooses are kept in place by blades of grass, and should have their lower edges about three inches from the ground. Over the cross-stick thick-leaved pine branches are placed, with snow to cover the whole and protect the nooses from the weather.

A simple kind of net for taking capercailzie, we read in L. Lloyd's "Game Birds of Norway and Sweden," is termed the "kasse," and can be used at any season of the year. It is about thirty inches square, and made of twisted silk with meshes so large as to readily admit the head of the bird. If there is snow in the forest the net should be white, but if the ground is bare, green or some other dark colour. This net is also hung across a cattle path, the four corners secured to bushes or to pine twigs inserted in the ground for the purpose, by means of woollen threads of just sufficient strength to maintain it in position. A stout silk line is passed through the outermost meshes of the net all round, and both ends secured to a neighbouring sapling. When the capercailzie gets his head into the meshes he rushes forward, the woollen threads are broken, the net drawn up into a purse-like form, and the bird rolls over helpless with his wings closely pressed together.

Then again, many a plump young bird that deserved a better ending has been cut off by the "stick-nat." This is a net usually sixty or seventy fathoms in length, twenty or thirty inches in depth, with the meshes some three inches square. The "telnar," answering to the cord and lead lines of our flue-net, consists of stout packthread, but instead of being fastened to the web itself they merely run through the outer meshes, and hence the net travels on them in like manner as a curtain on a brass rod. Stout sticks previously blackened by fire, and sharpened at the lower end for more ready insertion in the ground, are fixed crosswise to the net, or rather to the "telnar," ten or twelve feet apart. The "telnar" is about one-third shorter than the net itself, and consequently there is a quantity of loose netting called "los garn." On the net being set this loose netting is drawn up in folds to the cross sticks, and when the capercailzie runs into the net, the "los garn" forms a sort of bag about the bird, making escape next to impossible. The fowler takes his place in the centre of the netted circle, and by "lacking" —*i.e.* imitating the hen's cry—attracts and generally secures the whole of the covey of young birds on whose haunts he has placed the net after flushing them. The pullets only come to the pretended calling of the old birds when they are very young.

Then there are the blackgame—the russet hens and young cocks on the game-dealer's hooks being, perhaps, often mistaken by the careless for Scotch grouse, but the male bird is steel-blue, and white under-plumage is distinct, and not easily confounded with the lesser sorts.

It is from over the sea that our poulterers' shops obtain replenishments of blackgame. Podolia, Lithuania, Courland, Esthland, Volhynia and Ukraine, are all forest countries, and here indiscriminate shooting and snaring go on. During the winter season, in Siberia, they are taken abundantly in those elaborate set-traps which the traveller must have noticed along the sides of the roads between villages. A certain number of poles are laid horizontally on forked sticks in the open forests of birch; small bunches of corn are fixed to them by way of lure; and at a short distance off tall baskets of a conical figure placed with the broadest part uppermost; just within the mouth of the basket is set a small wheel, through which passes an axis so nicely fixed as to admit it to play very readily, and on the least touch to drop down and again recover its position. The birds are soon attracted by the corn on the horizontal poles, and after alighting upon them and feeding, they fly to the baskets and attempt to settle on their tops, when the wheel drops sideways and they fall headlong into the interior.

Every one should read those fascinating volumes of adventure in which Lloyd recounts his experiences amongst the Northern pine forests. His pictures of snow-covered trees, with a black-cock on every branch, eating the tender resinous shoots said to give their flesh a peculiar flavour at times, are enough to make the shooter envious indeed. The Russian peasants build huts full of loopholes, like forts, where the sawyers have been at work in the forests, or where an open glade presents an opportunity; and decoy birds—mere artificial imitations made of black cloth—are arranged around. As the grouse assemble the shooter fires through the openings, and if the sportsman succeed in keeping himself out of sight he may litter the ground with slain, as the birds are not frightened by the mere sound of the gun—a curious weakness of the grouse family. From Norway and Sweden many a box comes to Leadenhall Market, with this advantage, that less demand is made upon the tenants of our own highlands, and thus a plentiful supply is left to add variety to the "mixed bags" which are made before winter comes, and add point and interest to many a long day upon the "birken braes" that would never have been "trudged-out" but for the enthusiasm their pursuit arouses.

While black grouse killed by the score in these fashions sell at five shillings a brace, the "white grouse" of old writers, or, more familiarly, the ptarmigan, only reaches the modest figure of ninepence or a shilling each bird. But then they are pursued relentlessly. Greenlanders capture them in nooses hung on a long line, and drawn by two men, who drop the nooses over their necks. They eat them with train oil or lard, and their skins are converted into shirts to wear next the skin. Laplanders take them by forming a hedge with boughs of birch trees, leaving small openings at certain intervals, and hanging over each a snare. The birds are tempted to come and feed on birch tree catkins, and when they pass through the openings are caught by the neck and strangled. As "Bushman" says in "A Summer and Winter in Lapland": "The Laps select a birch sapling six feet long. It is cleared of twigs, and a horsehair noose fastened a little above the point, which is bent down and lightly stuck in the snow, the noose being about a hands-breadth above the surface. Small hedges are then built up either side, and catkins or fruit stuck on their thorns. When the bird walks up the avenue to get the bait he becomes entangled in the noose, and his struggles free the bent sapling, which then flies up and hoists him out of the way of foxes, wolves, etc."

On the Hudson's Bay territories also nets twenty feet

square are used for the capture of ptarmigan, and they are so numerous that ten thousand have been taken during a single season lasting from November to April.

In reference to the former snare, and showing how the same idea occurs to different people, it may be mentioned that the Aleut Indians of Canada use snares of twisted deer sinew made into a running loop and attached to a pole nicely balanced between two branches, the noose end held down by means of a small pin tied to the snare. Rushes are then piled on each side of the tracks in which the grouse run, so that they have to pass through openings in which the snares are set; a touch loosens the pin, and the heavier end of the pole falls, hanging the bird in the air. Probably the willow grouse is here spoken of—a bird that is not uncommon in the London shops, and very numerous in the Canadian "backwoods."

An allied bird shows equally little appreciation of danger. A writer on caribou hunting in *Forest and Stream*, says: "Just after crossing Murray's Brook, passing through some heavy timber, we flushed from the trail a spruce partridge, which alighted on a limb about eight feet from the ground. William was at first going to throw his axe at it, but Joseph urged him to snare it. A pole was cut and trimmed, and a noose made from a bit of salmon twine tied to the end of it. While this was being done, the simple little bird sat cuddled up on the limb, unconscious of danger, not even looking at us. When all was ready, William took the pole, and stepping quietly up to the tree passed the noose over its head, and dragged the innocent fowl from its perch." This process repeated several times, always with success, would seem to owe its practicability to the tameness of game inhabiting a region where human footsteps rarely penetrate.

I have been led somewhat far afield, and fear space will only admit of a glance at the "dodges" by which red grouse are trapped, to the chagrin of honest sportsmen, and the spoiling of not a few shootings. Tramps and loafers from the nearest towns set horse-hair nooses in their runs amongst the heather; and, as they often forget where many a springe has been placed, such nooses may remain set until the spring, and take parent birds, with nests adjacent. Again, in autumn, the reapers—always ready for a little work of the kind—stroll out of an evening, under the pleasant yellow harvest moon, and peg down fine nooses atop of the barley shocks. As a result of this sundry grouse are found there, flapping helplessly, head downwards, next morning, before the mists are off the low meadows. A few find their way into rabbit traps, Yarrell tells us, set on open moors, and one has been taken in a steel hawk trap, on top of a pole; but of all destructive and objectionable methods, netting of grouse by fixed nets is, perhaps, the worse.

They consist of long lines of fine netting hung on poles, usually by the proprietors of small, narrow allotments facing big moors, where a large head of game is reared. Now, when grouse fly, they rise ten or twelve feet perpendicularly, and then rush forward at the same height with a velocity which must be seen to be understood, and thus they plunge headlong into the meshes with a force which generally disables them at once. Instead of fair give-and-take, which is the rule amongst neighbouring landowners in matters of game, these daytime poachers take all they can lay hands on, and hardly rear a score of grouse, in return.

All this misplaced ingenuity is painful enough, and I turn with satisfaction to more legitimate manners of sport, adding a sketch or two from my note-books of quiet days upon the heather and solitary scrambles amongst pine barrens, dear to the naturalist as well as the sportsman. If I succeed in beguiling an idle half hour, as my own half hours have often been beguiled, by classic pens in the literature of out of doors, or in recalling pleasant remembrances, the object of these chapters will have been fully obtained.

### "The Twelfth."

But "the twelfth" is the white letter day of the heather trudger. He may enjoy it "in the ranks" of a noisy but well meaning party of lowlanders or alone. On a recent occasion—various circumstances prevent us, however, from doing conventional justice to the occasion by a big muster of guns and a proper day's shooting—myself and J——, equally enthusiastic, determined to try our fortune alone since we could get no one else to join us.

Be the party big or small, the weather is always a matter of the first importance. Fortunately it was fine and bright when we turned out at seven a.m. on the morning of the 12th. The sun was just rising behind the hills on which the old Scotch house was built, and throwing clear blue shadows of pine-clad summits half way up the opposite side of the valley, where the land was long heather and patches of coarse grasses, broken up by thin mountain torrents and veined by grey stone dykes. This was promising enough, but from some cause—perhaps the purity of the air—the sky in the Highlands is almost always blue and clear at sunrise, a state of things which early rising but inexperienced southerners take to be a sure token of a lovely day. Unfortunately the promise is often broken. This time, however, a fair sky was accompanied by a strong hoar frost covering the grass in the shadows of the trees with a beautiful powdering of white crystals, and glittering as it melted into dew-drops in the fast-increasing warmth.

I roused up the other "gun" who was to accompany me, and by 7.30 we were hard at work at breakfast, dividing our time between the hot coffee and many good things, and getting into our "war-paint." We agreed not to trouble ourselves with any dogs, and when I suggested that a keeper should come with us, J——, who is rather a Philistine in such matters, said, "Oh, bother keepers; let's see what we can do quite by ourselves! We shall be much freer to talk and smoke, and if we shoot more game than we can carry we can easily 'cairn' it for the moment." So it was settled, and with guns on our shoulders and bags at our back we started as the blue reek, beginning to ascend in thin columns from the many chimney-stacks of the old lodge, told us the household was astir.

Loading up at once, for there were large woods all round the house, and my companion declared his intention of shooting everything he saw—feather or fur—we entered one of these pine coppices and speedily found the rabbits had not yet retired for the day. The first shot fell to me, and a rabbit was bowled over as he bolted across the path ahead. Then J—— scored a right and left in good style, followed by three or four more as we went forward. Every now and then we had to stop to admire one of the many ferny hollows amongst the grey rocks and under the drooping branches of the spruce firs; dingles so deep and shady that, even at midday, they were silent and cool, and the sunlight only made its way through the thick roof of leaves overhead to play about for a short hour at midday on the soft carpet of moss and short grasses.

At the top of the wood was a shallow pond—a tank we should call it in India—of an acre or so in extent, and much overgrown with rushes. This we approached with caution, but no sooner did the smallest patches of our deer-stalking caps show through the bushes than the ducks we had expected to find "at home" rose with loud splashing and many guttural quacks. I had one chance for half a moment at the mallard as he went off through the tree-tops, and firing in an instinctive manner, the moment the gun touched my shoulder, he fell back headlong into the water, and, after a desperate endeavour to dive, succumbed. My shot put up a family of teal from the far end of the pond, who immediately separated and flew round and round their home as though loth to leave. This hesitation was fatal to one of them, for I got another snap-shot exactly as the bird came between myself and the sun; and, very considerably to my astonishment, down he came on to dry land. When we went to pick him up, the sunlight on his extended wings and head really made him appear as lovely a bird as there could be.

A little further on a heron flew overhead, out of shot, however—like all his kindred, shy and careful. I remember, one morning early, rowing up a quiet and secluded creek at the estuary of a Cornish river, and as the tide ebbed, I punted the light skiff along by the shallow margins, and no less than ten times got successfully within sixty yards of as many separate herons, all busy fishing; but nothing would persuade them to let me come just the requisite fifteen paces closer. The manner in which they rose and flew away, directly that distance was passed, was most striking, and showed a wonderful unanimity in their ideas of safety. These Scottish herons often breed on the ground, on rough mountain sides, contrary as it may seem to their general habits.

Finally, after a long pull up-hill, we came to the outskirts of the moorland, and "forming in line," as J—— said, we proceeded to work at once, for already the sun was fairly high, and the grouse might be expected to have made their early morning meal of heather tops—rather dry food, one would fancy—and to have settled down for a comfortable siesta on the sunny side of the grey boulders, or heather clumps, stretching as far as the eye could see.

With what a rough interruption the 12th comes to these pleasant morning meditations of the grouse! What a panic must seize the new broods, and how the old birds' hearts must fail them when they hear the guns, and know the great anti-grouse conspiracy of last year has broken out again! Our plan was to walk about fifteen yards apart in long beats across the range on which our moor lay; so working our way up to a well-known summit, almost amongst the clouds, where we might find a wide prospect to gaze on as we made our midday lunch, already the faint reports of the guns were coming from all sides, as the day's work commenced on the neighbouring moors. Across the valley, where the land, rising abruptly, exposed all the face of the deep-tinted strath to us, we could see a party at work, and in the bright morning air, as thin and limpid as ether, it was easy to recognize several well-known forms of friends and broad-shouldered gillies. Now and again we caught a faint glint of sunshine from a gun-barrel, and then there would be a puff of cotton-white smoke, followed rapidly by another and another; and as the wreaths lengthened out on the light breeze the sound of the shots came to us one by one, and perhaps even the shrill whistling of a keeper, calling in a wild young dog that had gone in pursuit of the covey up the hillside.

The first thing we put up was an old cock grouse, who hardly showed for a moment, as he went down a hollow; but this fixed our attention on the work in hand, and we went forward with guns ready and determination on our faces. Up got another bird, and, determined to draw first blood, J—— fired immediately. We walked up, and there lay a grey-hen—a forbidden bird until the twentieth. However, J—— said nothing, but slipped it into his bag, and we moved forward. Two or three grouse followed, one at a time, and we are just beginning to keep rather a lax look-out—my companion marching along with his gun over his shoulder, and myself lost in admiration of the wide-stretching ranges of mountains, rising to the northward step above step in tiers to the sky, purple in the shadows, green in the sunlight, with twenty shades of grey blending above—when from the long heather at our feet comes a cackle, a flapping of wings, and up rise some fifteen grouse as though they had all been thrown up by the same spring. We got one with our first two shots, and another with our next two—by no means good shooting, but, to tell the truth, we did not expect them just then, as my companion said. We picked up the slain, and as we straightened down their feathers, could not help admiring their beautiful sleek forms, their roundness of shape and compactness of build; in fact, one could almost trace the effect of the healthy mountain air in them. They would no more suit the lowlands than a loch trout would become an Essex ditch.

A few yards further again a solitary bird got up on my side, and was brought down in better style. Then J—— had a chance at two with like result, and so we went along for a couple of hours. Whether it was that the birds had not done feeding, or for some other reason, they lay well, and in general rose by twos and threes instead of coveys, an arrangement which suited us well, as the wholesale rises on this, the opening day, shook our nerves very considerably. I have listened to wild elephants charging through the dense bamboo thickets of a southern Indian jungle, and expecting every moment to see a great colossus bearing down on my stand, but somehow it was not half so deranging to my shooting as the sudden springing from heather of a whole concourse of loud-winged grouse. A little later on, when the bags were becoming very heavy and our thoughts turned to lunch and the bottled beer waiting for us, we entered a rough piece of land with a rather thick growth of spruce firs. This we beat carefully, until about the centre our nerves were again tried by the rising of a monstrous brown bird, which, as it went away down the slope like a runaway boulder, seemed as large as a big turkey. We both threw up our guns, but J—— fired first, and down it came amid a cloud of feathers, though the distance was a fair forty yards, and the shot only No. 6. We knew it must be a capercailzie, and such it turned out to be, a fine young bird of nine or ten pounds, an addition to the bag which decided us at once to strike straight up the hillside to the spring, at the edge of which we were to tiffin.

There seems to be no legal close time for these grand wild fowl; as far as Scotland is concerned their protection may safely be left to the owners of the wild mountain forests in which they dwell. They are not much in the poacher's line, and with fair play from the sportsman will probably be well able to take care of themselves. As to the common charge brought against them, the damage they do in pulling off the young tips of the spruces and firs is as nothing to the havoc made amongst those trees in the same way by the squirrels; but even were they very guilty on that count, they are magnificent birds and well worth a little indulgence.

We had been at lunch some few minutes, comfortably seated in the long heather, with the provender spread out in front and our guns behind, when a demand rose for water to mix with some "whuskey" which we proposed to drink to the success of our sport, and going down to fetch it from the little pool that bubbled up close at our feet, J—— put up a woodcock from the stones where it had been crouching and watching us, certainly not twenty paces away. It was useless for my companion to call to me to fire, for by the time I had got my gun the "cock" was half way across the valley.

Then our lunch proceeded in peace, and for a time we divided our attention between æsthetic admiration of the glorious wide prospect stretching around us in an amphitheatre of rugged hills, broken here and there by pale mountain tarns or rushing streamlets, and the more practical occupation of demolishing beef sandwiches and emptying sundry bottles of beer. It was curious to listen to the silence which had come over the valley; every one seemed to be at tiffin like ourselves, all the guns were hushed, and nothing broke the stillness but the occasional call of a grouse down below getting his scattered family together, or the far-heard whistle of a curlew.

We spent half an hour over the after-tiffin pipe, and then rather reluctantly roused ourselves, stretched, and after having cairned the game and the luncheon basket with heather and rocks, we shouldered arms and again proceeded to carry the war into the enemy's country.

The afternoon added a few brace to our total, a hare,

which was very cleverly stopped by J—— at a wonderful distance, a plover that flew overhead in an irresistibly tempting way, and a couple of wood pigeons returning from a foray in the low-lying barley fields. We were also guilty of the lives of two hen blackgame, which met their fate by rising amongst a covey of their cousins the red grouse, and under such circumstances, in the hurry of the moment, I for one can rarely tell the difference between the two species.

Finally, as the sun sloped down in the west and the grey rocks were beginning to have very distinct shadows, we reached the outskirts of a deep pine forest, clothing the whole summit and side of a hill on our ground, where, bidding adieu for the time to the grouse, we scrambled over the lichen-grown boundary dyke, and picked our way amongst the dense stunted firs, all on the *qui vive* for another capercailzie, several families of which made this their special home. The intense solitude and wildness of these "pine barrens" are difficult to describe to one who does not know them well. For my part, when I first made their acquaintance fresh from the lowlands, I was struck with surprise. It seemed some chance had swept me from the crowded little British isle to the wilds of Siberia. And what bird befits these great solitudes so well as the hermit-like "cock of the woods!" But I may have something more to say about these grand game birds below. For the present we made one beat lengthways through the dense forest, already in the gloom of approaching dusk, adding to the bag, a woodcock, another capercailzie, a couple of wood pigeons, a hare, and half a dozen mountain rabbits—a species differing very considerably from the lowland form.

When we counted up the spoil that evening in the verandah of the lodge, though not large it was pleasantly varied. We had stocked the game-room for the time, and if we had not shot very many brace of grouse, we had at least got that what we went out for—a capital rough day's shooting.

## Among the Blackcock.

Why Blackcock enjoy exemption from the dread disease which ever and anon carries off their near relatives, the red grouse, sends down the rents of highland shooting, clipping the expenditure of the lairds, and influencing the finances of one-half of the British isles, can only be determined when the true nature of the malady is better understood. But many things favour a long and happy life to these birds. Their food is various; frost-bitten heather in the spring matters little to them; from the time the snow melts on the mountain tops to the period of its coming again, their bill of fare is ample and full—thus they are placed beyond the reach of hard times and sickness-bringing scarcity. Then their powerful bodies and large size must modify the courage of attacking hawks, giving them little "stomach for the fight," the same cause doubtless repelling the attacks of the marauding hill crows and ground vermin, who think twice before robbing the nest of so stoutly made a bird as the watchful and courageous grey-hen. Lastly, it might be suggested that the freer flight and less gregarious habits of the moor-fowl save them from what is probably the most pregnant cause of grouse disease, the overstocking of estates.

This preamble, however, is merely to introduce the fact that, disappointed with the grouse last season, and yet bent upon getting something in the way of sport out of our ten thousand acres, I turned my attention, directly the 20th of August gave lawful sanction, to the "grouse of the second degree." Let me place before you the surroundings of myself and another equally ardent "gun" on the eve of that long-looked-for date. No palatial shooting-lodge this time. A scorn of rheumatism and a taste for roughing it had determined us to take time by the forelock, and to march out into the enemy's country over night and camp in a shepherd's "shiel," in order to find the birds on the feed the next morning by dawn, and run up half-a-dozen brace if possible before the stay-at-homes were even thinking of turning out. The time, then, is eleven p.m.; we have made our way three miles out to our destination, and a roaring fire of birch logs flashes and crackles in one corner of a rough stone hut of very modest dimensions; the grey smoke ascending in spirals to the roof of heather and bracken fern, whence, after much consideration and many contortions, it finds a way through a weak corner, and disappears into the darkness. Though rough, the hut is by no means uncomfortable. The crannies between the stones have been filled with moss and fern, while plenty of both at one end of the cabin form a delightful lounge, either to sit or to sleep on. The guns, cartridge bags, etc., with a stray head or two of game picked upon the way out, hang from pegs or lean in corners, while my companion heaps logs on the fire with one hand, the other meanwhile keeping in scientific motion a frying pan, whence comes a most appetizing odour of grilled supper. I myself, having fetched an ample supply of water from the neighbouring burn, demand and obtain a place for the kettle on the fire, when a brew of tea is soon ready, and in less than a minute we are hard at work at our simple meal, our knees for tables, and wide rounds of home-made bread for plates. At such times the conclusion comes with irresistible force, that too much culture deadens half the enjoyment of life, and that man in a state of semi-wildness, "earning the food he ate, and pleased with what he got," must indeed have lived in the true Golden time. Perhaps more mature consideration will lessen the envy with which a man is apt to regard such a state of simplicity, for it is a very doubtful point whether freedom from butchers' bills would compensate for an occasional involuntary fast of a day or two when game was wild or scarce. Yet a return now and then to primitive manners, an unshackling of the harness of civilization, and a brief period spent in imitation of our huntsmen ancestors, must always prove attractive to a well-constituted sportsman's mind and body. The camp fire alone is a delight of the first degree. "Men scarcely know how beautiful flame is," nor truly appreciate it; but when a thousand stars are twinkling overhead, the crisp crackle of the wood and the flying sparks impress the mind with a pleasant sense of companionship in solitude and love for the great Promethean gift which is felt but dimly under more familiar circumstances. Yet, though Lares and Penates are usually looked upon as strictly household gods, the resting-place of the shooter or traveller, if only for a short time, needs its altar as much as does the most fixed abode. What could strike a pilgrim with more sense of discomfort than a halt under the canopy of heaven without the cheerful light of leaping red flames? Can we imagine bright stories and laughter as the evening meal is made among men sitting with feet towards darkness? No; the idea is barbarous. Little matter place, time, or temperature, the sojourner in the wilds, on halting, turns his first attention to a cheerful fire, in presence of which he can contentedly enjoy well-earned repose. With it the hunter's food is ambrosial, his drink, though it come from a hill stream, is nectar for the gods, while his sleep, if it be only on the mattress of earth, is the choicest gift in the liberal apron of good mother Nature.

All these pleasures, which make out-of-door life so fascinating, we acknowledged as we sat by our fire, enlivening the evening by stories and laughter, and piling up the logs till the flames threaten destruction to the roof of sod and heather, our only protection from the night dews; till, our pipes having burnt out twice or thrice, we drank health to the morrow, and, wrapping ourselves in the ready tartans of our adopted heath, with a final touch to our heather couches, were soon in the unsubstantial hunting grounds of sleep. But the pleasantest repose will give way before a prearranged determination to wake at a certain hour; and thus the earliest dawn, stealing through the chinks of the doorway, disturbed us as effectually as a louder summons would have done. We were soon up, and while the other gun replenished the camp fire I went for water from the tumbling stream to make the early coffee, the very thought of which gave us an appetite. How fascinating the world was in its "beauty sleep!" The sky an undecided purple, with here and there a star twinkling faintly; and, down in the east, a great straw-coloured planet lying just upon the deep, black, rocky outline of a towering mountain summit. The stillness meanwhile was worth listening to. Even the rill by which I stood, regardless of my errand, seemed quieter than usual, and fell into its deep pool between the rocks less obtrusively than heretofore, not another sound breaking the silence far or near. The whole glen, indeed, was buried in calm repose and peace; below, the black, profound, silent shadows, contrasted here and there with pale streamers and patches of mist marking the bogs or peat holes; above, on either hand, against the sky the rugged edges of the hills were now just touched with a suspicion of the coming day, their outlines growing sharper every minute. But an impatient shout from my companion brought me to the contemplation of the practical. The kettle was soon filled at the bubbling cascade, and, hurrying back, we were forthwith busy in the preparation of a hasty meal, for we were bent on watching the sun make his rise from a point of vantage, and there was little time to be lost.

Nor was our energy without its reward. The meal over, and the things replaced for the moment in the hut, with guns on our shoulders and our sprightly dogs at heel, we boldly turned our faces to the steep northern ascent; and, hand over hand, through deep rock-bestrewn bracken, and dim ghostly tangles of dwarf birches and alders, silent and quiet in the cool air of the early morning, we made our way, until, somewhat breathless and warm after ten minutes' hard climbing, a rocky ledge was gained commanding a magnificent panorama, and we sank down on the moss to await the rising of the sun.

Soon my companion exclaimed, "Here he comes!" as a pink light flushed into the sky like the reflection of a far-away conflagration. Quickly it rolled up till it reached the purple sky right above us, deepening into crimson, and bringing out with a touch of colour on each, that grew every moment more vivid, squadrons of light fleecy clouds, of whose presence we had hitherto been ignorant. Then the light in the east streamed strongly up from the nether world, catching the distant hill-tops, and spreading rapidly from peak to peak, touching, as it went, each with gold, until they stood out from the purple shadows like the jewelled bosses of a shield; while the wonderful refulgence ran down the gullies, and, glancing from the high plateaus, passed, above and below, through a hundred changing shades of flame and orange.

Finally, while we were still watching this shifting transformation scene, before we knew it the sun himself shone from the brilliant masses of clouds, and all the hillsides woke to life.

For some time longer we sat in silence, admiring the beauty of the scene and the fresh, sweet air; but our thoughts soon turned to the object of the expedition, and being on likely ground, we at once proceeded in search of sport.

It has always appeared to me that the blackcock is a very early bird; to shoot him the start cannot be made too soon after sunrise. He rarely rises so well or seems so active later in the day, differing in this from the grouse; and should the sportsman wish to find the birds easily, and to see them on the move in all directions, he must adapt himself to their ideas of "catching the early worm."

No sooner are we started, and the spaniels "hied" on, than they begin, after a few casts to right and left, to draw ahead, with tails swinging nervously, and noses sniffing the ground. My companion nods significantly to me, and we close up with a dog, who, giving a hasty glance in the most sagacious manner, to assure himself that we are at hand, plunges forward, and out of a clump of bracken hurtles into the air a large bird, all black, who, with noisy wings, shoots fifteen or twenty feet upwards, and makes off up at the glen at a great pace.

We recover our composure as rapidly as may be, and I take the bird as he tops a stunted birch thirty yards off, listening with satisfaction to the heavy thud of his fall, when a motion of the hands sends a dog off at a gallop to retrieve him. We are following, when another cock gets up, and, rising high, tries to fly over us towards the opposite side of the valley; but this is the height of rashness, for we have already had eight days at the grouse, and are "in the swing." The other gun takes the shot, and the big bird comes down back first, with a long trail of feathers behind him. We cannot help admiring them for a couple of minutes. Mine is wanting a feather or two of his neck—a common occurrence at this time of the year, when the moulting season is on; but the other is quite perfect, and, as the first of the season, his twisted lyre-like tail has been promised to grace a highland bonnet on a certain fair Saxon head. The blue gleams of light on the back contrast beautifully with the delicate white of the slender feathers under the wings, the exposing of which as he rises makes him so conspicuous a mark against the green of the bracken ferns; but, to my mind, the finest thing about him is the bold build of his head—the strong black bill, slightly hooked and sharp edged, the thick neck set with glossy black feathers, and the bright eyes, with their curious overlay of close scarlet wattles, giving him a bold domineering expression that fits well with his disposition and habitat.

In size there can be no comparison between the lordly blackgame (the cocks of which reach as much as four or four and a half pounds, and the hens over two) and the smaller and lighter grouse, nor are they alike in habits. The grouse is a bird of great attachment to its mate. If, unlike the eagle, he does not remain faithful to her until death decrees a divorce, he yet keeps troth for a year and a day, doing some share of domestic duties, and taking part in beating back the pillaging hawks when they swoop down on the young broods. The blackcock is a roisterer of different habits, with affections so unstable that they only serve to make him "daft" and contemptible to all respectable birddom for a few weeks in the spring. It is he who comes down in the earliest mornings of the new year from his perch among the pine branches where he retires overnight, to be out of the way of prowling vermin, and to keep his body—of which he is very careful (the result of being a bachelor nine months out of the twelve)—out of the cold; and, winging his way through the thin mists of early dawn to some quiet open spot, alights, and commences that ridiculous love dance that has been so often described by naturalists and sportsmen. How any reasonable gray-hen can admire such a strutting, puffed up, and excitable wooer as he then shows himself to be, it is difficult to understand; but doubtless she knows it is all for her sake, and that, in the female mind, is excuse broad enough, no doubt, to cover any folly.

There every morning the cocks strut and crow, pacing round in well-worn circles with every variety of style; now and then fighting terrible combats with glossy black-armoured rivals, who come at their challenge from other ridges and slopes, and carry on the conflict before

> Store of ladies whose bright eyes
> Rain influence, and adjudge the prize.

But as soon as the frosts of winter have grown thinner on the hill-tops, and no longer, even at earliest dawn, turn to ice beads the dew on the burn-side bents, the blackcock retires to sober bachelor life, and for the rest of the year attends strictly to his own affairs; in pleasant weather haunting the highest ground he can find, and roaming hither and thither with a few other "good fellows" on the light wings of fancy; but coming down to the more sheltered hollows, where the hens assiduously sit or tend their chicks, when storms break above and grey mist sweeps backward and forward in a dull, damp sea of vapour along the mountain summits.

To-day, as soon as the sun was well up, we found the birds thickly upon the elevated ground we were now beating, which at another time, after a period of wind or rain, would have been useless for our purpose; but a little practice soon makes one familiar with such matters, and before long we brought ourselves to believe that we were as knowing judges of likely localities for the birds as they themselves were in selecting good feeding grounds.

Soon we approach a place where the land dips suddenly out of sight, obviously the deep bed of a mountain torrent, worn by countless ages of fretting; and here J—— makes a sign to me to approach with caution; so, waving back the dogs, who at once come to heel, we walk slowly to the brink and look over. Nothing! Yes, but there is! And down below us, perhaps fifty feet, are five blackcock on a little patch of green sward under a dead lightning-withered rowan bush.

For a moment or two, during which we are unnoticed, we watch the slow, leisurely way in which they are picking the seeds from the tall grass and rushes, and their self-satisfied air as they walk daintily about. It is a pretty sight, but very brief, for soon a bright eye is turned on us with doubt and hesitation for an instant, and then, when the danger in its full force bursts on the discoverer, and he recognizes the hated Saxon at arm's length, a hoarse cry escapes him, throwing the whole covey into a panic. With hardly a glance at the foe, they follow their leader's example, tossing themselves into the air and dashing off as fast as muscular wings can carry them. Forthwith our guns open fire, and, as the smoke clears away, a victim or two lie amongst the ferns and ling.

These are followed by others that we come upon suddenly,

still making their early morning meal in the soft ground among the sedges, or in pleasant alleys between the banks of bracken, just gaining its autumn tints of brown and amber —lovely enough, though somewhat melancholy, as marking the downward steps of the glorious summer.

We have as many varieties of shots as we could wish, and, in places where the broken rock masses are piled numerous and thickly, with overhanging brushwood, the shooting is very difficult. Now a bird will slip quietly off a ledge of such a pile of stones, and, gliding down hawklike with outstretched wings, will, unless we are very sharp, be out of sight before our gun can be brought to position.

Other birds will rise from among the thick tangle of vegetation and *débris* underneath fallen trees as the shooter approaches, stealing away on the far side with most aggravating expedition.

To show, however, how close the game will sometimes lie in such places, I may mention that on one occasion we came to a spot where some fallen timber was in confusion amongst the ferns under a clump of birches. We halted, and, not seeing any game in the neighbourhood, lit our pipes, and while resting for a few minutes, made, as usual, a fire, the smoke of which blew about in every direction; and yet, when we once more moved forward, our guns carried idly under our arms, up sprang a blackcock, followed by four of his boon companions, from a little island of bracken that we had looked upon with contempt. We were so astonished and taken aback that a couple of charges of shot sent after them did not touch a feather.

With such varied adventures—sometimes, by blunders or lost chances, going down deeply in our own estimation; and, again, soothing our ruffled spirits by a brilliant snapshot, or good piece of luck—the bag all the time grew heavier and heavier, until the finishing touch was put to our endurance by a gigantic blue hare, which, getting up between us, was fired at so exactly on the same moment that the two reports were merged in one, and he rolled over very dead into the dry basin of a little streamlet.

"I think you shot him," said my companion, dubiously lifting the heavy beast with some effort; but, remembering that we each carried our own game, I modestly tried to persuade him that it was his victim. But that would not do, so we forthwith built a hollow cairn of stones on a conspicuous ledge, and consigned our game to it until we could send a keeper up for them.

By this time we had worked our way to the crest of the range, and a fair prospect of hill and dale lay below us, a chequered plain of land and water as far as the eye could reach—lochs without end or number, so numerous they seemed, and at our feet the noble reaches of one made famous for ever by a touch of the magic wand of the great "Wizard of the North," "Loch Katrine's mirror blue."

"Let's try the locklet for teal," said the energetic gun at my elbow, for ever disturbing me when my attention is absorbed in the sublime; so we turned down a grassy slope on the plateau top, and, crossing some bare peat bogs, where the water, brown and dark, stood in the holes and ditches left by the subsidence of the surface, we walked in silence for half a mile, till a rugged hillock rose before us, and behind it lay an oft-visited mountain tarn, marking the water-shed. This spot was in general a sure find at this hour of the morning for teal or duck. We divided our force so as to take the enemy in front and rear.

Who is there that has seen one of these wild, unknown, unnamed sheets of water can forget the weird spot? More lonely places it would be hard to find upon the face of the earth. The hot sands of the desert, the dense, gloomy depth of a tropical jungle, never conveyed to my mind half the sense of loneliness that one of these little lakes does. All around their borders the gaunt, uncanny rushes wave and tremble as though at their roots lay some worse secret than that of the Asiatic king; and heavy, sodden mosses, green, yellow, and red as blood, stretch out on every side in a palpitating, aqueous flooring, fringing here and there unwholesome pools and dykes, where the water sits, wondering to what ocean it shall flow. Melancholy, prehistoric water-plants hug themselves with the idea that the world is back again in the Miocene period of its existence; and then, worse than all, killing the tender flowers, and ruling the region with endless tyranny, the mountain wind sweeps for ever over the morasses, chill and cutting even when the sun is at its highest, shaking the reeds and cotton grass, and ruffling the surface of the waters that lap perpetually with discontented mournfulness on the peaty margins of their prisons. Yet the ducks like such places, rearing their families in security, and we must suppose equal contentment, amongst the deep beds of rank water weeds. Here we hoped to find them; nor were we disappointed. Creeping round the sheltering knoll, and timing our walk so well that we both came in reach of the pool at the same moment, we examined its surface, and saw with great satisfaction a flight of widgeon riding in the centre on the miniature surf; some teal feeding on the mud with much satisfaction, if we might judge by their deep absorption; a brood of flappers under the care of an old duck, and a couple of mallards performing their morning toilets on a tufty island of coarse grass; in fact, our only wish was that there had been some more guns at hand to help in the foray. According to agreement, I crawled slowly forward again, after a minute's rest, in order to get as near as possible before they rose; but it is always the unexpected that happens. I had gone some distance down a rather wet peat channel, much marked with the "spoor" of sheep and mountain hares, till, thinking it might be as well to have another look at the locklet, I raised my head with the utmost caution, and was about to take a view of my surroundings, when a cluster of brown bodies in the stunted heather, not five yards away, caught my eyes; and there, close crouched, was a covey of red grouse, totally unconscious of my presence, but entirely absorbed in watching the movements of my companion a hundred paces off, who in his turn had both eyes fixed on the ducks, from whose sight he was well sheltered by a fallen rock. Such cases must often occur in the field. Every sportsman probably passes over much game that is well aware of his presence, though he may be totally ignorant of theirs; but it is not often that a third person gets a chance of witnessing unobserved the process. Needless to say, I was seen almost instantly, and the whole covey rose on the wing like one bird at the alarm cry of the old cock. The ducks also heard the cry, and, knowing by that curious freemasonry which exists amongst birds that it meant more than an ordinary summons to seek new feeding grounds, the "flappers" melted from sight into the sedges like shadows, while the widgeon and teal flew up, and, taking a wide circle, came directly over us with "loud whispering wings." J—— had already fired both barrels at the grouse, which he declared had gone by like a whirlwind not more than a dozen yards overhead, and had brought down (tell it not in Gath) three birds. So the widgeon were left to me, my first shot being an unexplainable miss, though the next one mended matters by stopping the hindermost of the flight just as he was passing out of reach. By the time we had reloaded, the teal, according to custom, came round again in a wide circle over the bog, and three of their number fell as they passed over us; but the mallards and other ducks had gone straight away down the valley.

Then we went down to the pond, where, after a brief bit of paddling, the dog came upon the brood of flappers, and put them up beautifully two at a time, and we got six out of eight with seven shots. By this time the sun was well up, and we were very conscious of the lightness of the early morning meal we had taken; so, distributing the game, we took a "bee-line" for the encampment, and twenty minutes afterwards we came in sight of the camp fire and a fine breakfast spread on the heather, already seated beside which were the other guns, who were to join us in the serious work of the day, after some much-needed refreshment had been taken. Many and various were the jokes attempted at our cost, but we treated them with the lofty derision we could so well afford, and never, not even late in the afternoon, when we were conscious of a certain stiffness about the knees, the result of early rising, did we regret the night in the open and the marvellous beauty of a highland sunrise. There is more sweetness in the early hours of this sea-begirt kingdom than perhaps one in a thousand of its inhabitants knows.

## Capercailzie Shooting.

Cunning and strong while alive, and by no means a bad table bird dead, the capercailzie lives amongst the finest of natural scenery, as we have said; to stalk and shoot him is fairly good sport with the additional attraction of glorious exercise. Driving the great grouse over previously hidden gunners is, however, little less than a shame. He does not lend himself kindly to this latter sport, and his bulk is so large that the simplest bungler who can pull a trigger gets more than a fair chance, as the mass of feathers, borne on broad wings, sweeps through the glades of the forest.

With this theory in mind, I on one occasion made a quiet raid upon the "cock of the woods" in his native fastnesses, before deeper snow than that already fallen on the hills round our Scotch lodge rendered his haunts inaccessible. Thus one morning, when all necessary preparations had been seen to overnight, cartridges loaded, boots greased, etc., we were ready for a start immediately an early breakfast was over—"we," on this occasion, being myself and a useful retriever, as fond of rough sport as his master, and possessing a keen nose, an admirable temper, and a thick coat, all essential requisites for the species of hunting we were going to undertake.

Forthwith we set out, climbing the wire fence that separated the civilization of the grounds from the wilderness of the woods beyond, and walking quickly over the crisp white snow, frozen as dry as sand by north winds blowing it hither and thither all night, until a shrubbery of pines undergrown by furze bushes was reached. Disregarding the rabbits that peopled this region and were skipping about amongst the roots in scores, I reserved my fire for a moment or two, as just ahead, at top of the little burn coming tinkling down the hill through a channel rugged with icicles, lay a reedy marsh surrounded by larches and overhung by willows—a likely spot for ducks on such a day as this; so we moved slowly up, taking advantage of thick patches of snow to deaden all sound of our footfalls, with increasing caution as we drew near the spot whence the surface of the ponds could be seen. A few yards further the willows rose above the gorge bushes ahead, and from the last sheltering bush the weed-grown surface of the partially frozen tarn could be observed. The first glance round was not promising, but a second and more careful scrutiny showed a bunch of ducks feeding quietly at the far end of the water.

Despatching a handy stable boy, watching the proceeding with vast interest from a neighbouring lane, to make a *detour* and take them in rear, I repressed the ardour of the dog, who was trembling in every limb with cold and excitement, and waited with eyes on the birds and finger on the triggers. For a few minutes they continued their methodical feeding, coasting along the half-frozen edges of the reeds, and now and again tipping themselves up to explore the mud of the bottom. But soon they get an uneasy fancy that something is approaching them from the far side, and up go their heads, and they crowd together, turning this way and that in nervousness, which comes to a climax as the form of our boy breaks through the bushes. A second afterwards, kicking the water into foam behind them, they rise as though lifted by one pair of wings, and bear straight down for me, sweeping over the pointed tops of the snow-laden firs, about thirty yards distant, with the early sunlight showing up the white feathers lining their wings. The leader gets a charge of No. 5, and comes down unmistakably to the dry ferns, and another bird has the other barrel, which results in it dropping both legs and falling in a long incline "grounds" about fifty yards away. This is too much for "Jack," who, with a yelp of delight breaks loose and returns in a few seconds with the mallard in his jaws.

Picking up the birds and slipping in two more cartridges, we go on again under the firs through a gap in a stone wall, and enter upon a tract of rather wild ground, where the rabbits were lying out in their snow couches in great numbers, to the intense delight of the dog, who chivied them hither and thither—bad form of course on his part and mine to allow it, but what can you expect from a dog who has not been out for a week? So I let him run riot for a time, but when ten minutes' tramping brought us to the slope of the great hill, and the long shadows of the pines fell on the snow above us, "Jack" was called up, and put in an appearance from a distant field with his tongue hanging out, panting prodigiously, and a general air of abashment. Matters were then pointed out to him, and he was instructed to restrain undue zeal and keep to heel, he at once taking up that position. We scrambled over a tumble-down stone dyke, and entered the pleasant shade of pine woods. What can be more lonely, yet what more attractive in its solitude to a lover of nature than a great pine barren? Once fairly in, the sky is only to be seen directly overhead. All round on every side, as the wanderer turns hither and thither, stretch the long silent vistas of the wood, scores and hundreds of fir stems, grey with lichens and long pendent mosses, stretching away to the remote parts, where they are blended into a confused mass that appears impenetrable until approached when the spaces open out, giving fresh views of new aisles. Here and there the monotony of grey is broken by the low thick branches of a spruce fir coming down to the ground, where they spread in an ever green canopy, forming snug hiding-places against chance showers; or perhaps one of these trees has been blown completely over, and, lying along the ground, forms just such a sort of shelter as the capercailzie loves.

Amidst such a forest of stems we found ourselves now, nothing to guide us to our direction but the slope of the land, which was, it must be confessed, very decided; and we were soon scrambling upwards hand over hand through broken masses of rocks, tumbled about like the ruins of a great city, the spaces between them filled up with deep snow, through which here and there appeared the tall stalk of a withered foxglove and masses of amber and golden fern. Scrambling over such stuff in the semi-twilight, with a heavy gun, a game bag, and supply of cartridges is decidedly warm work, tending to make the climber a little careless as to where he is going. Thus it was that the best chance of this morning was lost, a young roebuck upon which I came suddenly in a little natural hollow, vanishing almost as silently as a ghost before I could get my gun ready, leaving me not a memento but his spoor on the fresh snow, and the remembrance of his tawny hide as he glided down the valley. We did not pursue, being but too well aware of the uselessness of such a proceeding. These roebuck are most fascinating little deer, and many a bright summer morning when the blackcock have been calling and fighting all round, and the world has been wringing wet with dew, have I been after them. They are much harder to find than red deer, owing to close keeping to the shelter of coppices and forest glades, where a chance shot is all that can be got now and again. The only sure way of obtaining a shot is to lie up outside a plantation, long before dawn, and wait patiently for their coming out to feed; and they won't do that if you "stand between the wind and their nobility," or if they catch the smallest sound or the faintest movement from behind your screen.

However, to return to our capercailzie. After the misadventure with the buck, strict attention to business was the order of the day, and being so high up the mountain side that a vast extent of snow-laden fir-tops were visible below, I struck along the slope—decidedly better walking—and proceeded with due caution. Everywhere round about the white covering of the ground was pitted with marks of the mountain rabbits which abound in these wilds, and were skipping hither and thither in tempting style, which would certainly have brought retribution on them had I not been after better game. These hill conies are as different as can be from their cousins of the lowlands; their fur is much greyer—more like that of the badger, their limbs are shorter, and their build altogether closer and more compact. It might have been feared that naturalists, in bestowing Latin names on the group, would have taken note of these facts and made the variety into a species; but it is well it has not chanced so, for such dividing where Nature has made no division is not to be commended—there are already only too many instances of it.

A minute or two and a fallen pine tree appears lying in a vast mass of green confusion across the rocks a little above us. "Just the place!" I mutter, and scramble towards it with gun ready this time, and then pause about fifteen yards off. A moment of silence succeeds, and the dog is on the point of being sent in, when a mighty flutter takes place spontaneously, and a brown mass "quits" on the far side, but keeps the tangle of branches so cleverly between us that I am quite unable to get a fair sight, and is away through the wood in a second. There is no time for the sorrowful reflections which might else have followed, for another prodigious disturbance occurs among the partially withered branches, and amidst a cloud of disturbed snow the cock of the woods himself rises boldly from the midst of his harem, shooting up for the tree tops above with a speed and ease wonderful in such a bird. It is a fair chance. I follow his flight for a moment, and then with a crack the report resounds through the wood, followed after a moment by a heavy rush from above, and a thump on the ground. We rush up, and there lies the gigantic capercailzie in his final struggles on the beaten snow, which he is tinging with a crimson stain. I feel reasonably proud, as he is in the best of condition, and will make a display when he is got home. But that is a somewhat arduous task. He is about the size of a medium fat turkey, much too big for the game-bag; so his feet are tied together, and he is slung behind, where he rides comfortably enough for the time being.

Then on again to pick up another, if possible. But now, the ice having been broken, the dog is sent a-hunting for whatever he may find, and I am all ready for the next chance, which comes pretty soon in the form of a nimble mountain rabbit springing from a shelter of fir branch on the ground, and making off up hill closely followed by that graceless dog of mine.

However, it distances him in a yard or two, getting out of sight for a moment among the boulders, appears again higher up the hillside, where a sharp shot stops it all of a sudden ; its four-footed pursuer running in and proceeding to mumble it in a way that earns him a well-deserved reproof. After all the sound of a gun is out of place in these solitudes. The report, shut in by the close barriers on either side, echoes and re-echoes in a startling manner on every hand. Were the heavy sound which desecrates the hillside to call forth some monstrous shape or old world vision, one would hardly be surprised, so ancient and solemn is this abode of silence, with the long tawny lichens hanging in ghostly lacework from the warped and stunted firs and shattered rocks, rolled down from above, like disjointed masonry, taking strange shapes of turrets and witches' caves under the thick canopy of flat branches and dead bracken. Here it is said, indeed, there does actually dwell a harpy of evil form, whose chosen home is this stretch of gloomy wilderness, and some imaginative bard has turned, for the benefit of wonder-loving southern tourists, a distich, whose intention is better than its metre—

> "If the fiend ye offend of the knock of Balmyle,
> Your life shall you live but a very short while."

Many a time I have hunted in the territories of this being; watched a sunny vale below from the summit of his chance-piled castles; lopped branches from his oldest trees; lit fires in his deepest caverns, and inquisitively penetrated his densest tangles! In fine weather, when the golden sunlight is speckling the floor of pine needles with patches of shifting colour, the tall foxgloves rocking in the gentle currents of air sighing on the tree tops and loaded with the faint aroma of sweet-scented resin, while the soft notes of the shy wood-lark or ever active goldencrest have been the only sounds to break the stillness, the pursuit has been one of endless amusement. But it is another thing hunting a titular family demon towards the close of a short winter day, when every unseen rivulet chafes angrily in its bed loaded with blood-red peat water, and the firs lash about their rough arms, tossing them up to the cold rising wind, and creaking above and below like the scantlings of a ship in a cross sea. At such times the curlews, bound southwards, sweep overhead with unearthly cries, and the mist comes down deep and sombre, hanging about among the rocks whose weird shapes are more than ever fantastic through its dim folds. If you have ever listened in a shepherd's cot to wild highland tales of superstition; if you have ever had even a suspicion of a belief in ghosts, this is the time that you will, in spite of your best efforts to put such fancies behind you, think of everything gruesome you can remember.

Indeed, these forests of the Scotch highlands are not to be "sneezed at" for that half of the year, when not one lowlander in fifty, of those who shoot through the short glory of a Scottish autumn, knows anything of them.

One such sportsman said to me once, after we had emerged from a short cut down a belt of forest on the edge of steep corries, where the mist was lying pretty thick, and the rain passing in squalls over the tree tops, that he had never seen such an "infernal region in his life; and as for banshees, why, it is just the most promising cover you could want." I was bound to confess I had never come across anything more weird than this locality, in many wild expeditions in both hemispheres.

The highland rabbits like the hillsides; the broken ground gives them great protection from beasts and birds of prey, and they do but little digging—"the conies are but a feeble folk, yet they build their burrows in the rocks" applies here very well. Yet, even with this protection close at hand, they suffer occasionally from the talons of the free-booter. On one occasion I found on a hollow ledge of stone, protected from the wind and rain by an overhanging cornice, as many as five full-grown rabbits, every one with its back broken and both eyes pecked out, but otherwise untorn or uninjured, the work probably of a young and over-fed eagle, for one had been seen sweeping about the neighbourhood a few days before. The royal bird is clearly an epicure, or mighty fond of hunting! Owls, too, play havoc with the young ones, stealing along in the twilight and picking them off the hillocks where they congregate, with the utmost speed and certainty. For this reason, as for many others, the bird should be protected by farmers, not only in the north but in the south.

Snap shots were now the order of the day, and very pretty shooting it was too, turning the bunnies up from their warm shelters under piles of withered fern fronds, and taking them "on the hop" as they dodged between their abundant covers.

Twenty minutes of this sort of thing satisfied me for the

day, and it was a relief to drag a bunch of mountain bunnies into the nearest ride, where they were strung up by their heels along with the capercailzie on a strong branch, there to await a keeper's boy sent to fetch them home, and not forgetting a scrap of paper torn from a note-book and fixed conspicuously to them for the double purpose of marking their whereabouts and scaring away any prowling lynx-eyed corbie-crows, who like nothing better than a share of another man's meat taken on the sly.

Eschewing fur, the next addition to the bag was a brace of wood pigeon returning to roost, whose shadows gliding across the snow at a spot where the ground was bare thus betrayed their approach, one falling to the first shot, and the other of these birds, who would seem to keep constantly in pairs all the year, as she came wheeling round to see what had happened to the other, sharing a like fate. The wood pigeon is one of the loosest-feathered birds existing. The spot where they fall is almost invariably marked by a perfect litter of cast plumage, and no other bird get so draggled and spoiled in the republic of the game bag as they.

Lower down, where a mountain torrent spread itself out over a land delta of its own making, in a number of thin streams, a woodcock got up and sped down the glade with hawk-like speed. This was too pretty a trophy to be lost, and so a charge of No. 6 at forty yards brought the russet plumage to the ground. Another was shot some way further along the slope; but though these two had roused my enthusiasm, and more likely ground was beaten under the lea of the wood, no more were put up, and being now on the low ground again, with the firs towering tier upon tier overhead, I came to the conclusion that the capercailzie should have a rest for the present, though no brilliant score had been made no doubt, and turned my steps homewards. Just at the edge of the plantations fringing the roadway leading to the house, a cock pheasant strutted out of a ditch, and finding himself in unpleasant proximity to the dog, took a short run, a couple of hops, and then launched himself into the air with his tail streaming gallantly behind him. Thirty yards' law given him, and he gets a dose of "leaden hail" that brings this brilliant game down to the snow, and this is the last shot of the day.

The guests are out sleighing, but when they return to lunch there is a row of birds waiting their criticism on the grass by the porch, two capercailzie, a cock and a hen, the latter shot on the way home, nine rabbits in their thick grey winter fur, a handsome mallard, two wood pigeons, ditto woodcock, and last but not least the pheasant, which the keeper's boy strokes "gingerly" from its glossy green-head to the tip of its long unruffled tail, before he places it at the end of the line. Such is the sort of mixed bag it is possible to make when the snow covers the Perthshire ranges, and regular sport on the heather or stubble is impracticable.

## The Last of the Grouse.

"Can you drive over here for a final harrying of the birds to-morrow, before we go south?" wrote the son of a neighbouring laird a short time ago, and knowing the invitation would be backed by pleasant company and at least fair sport, and that of the kind which, late in the season, is practised on most estates, I most willingly sent back an acceptance.

Looking out the following morning the prospect was wintry enough. All the higher spurs of the ragged neighbouring mountains lay shrouded in snow, where a few hours before they had been green and fertile. Truly the hand of winter was coming down upon the land, and in a little time even the few still occupied shooting lodges would be bare and empty of their summer migrants. But we judge things as they affect ourselves, and the snow would make little difference to-day, since it was confined to the higher ranges, while our working ground for the time would be on comparatively low-lying moorland.

Breakfast over, myself and J—— climbed into the waiting dog-cart, in which guns with cartridges *quantum suf.* were ready stowed away, and tucking in the comfortable rugs, for an autumn morning in the Highlands before the sun is well over the hill-tops is none too warm, J—— picked up the ribbons, flicked the sleek-coated chestnut, and away we went down the drive, our cigars aglow, and minds full of pleasant anticipations.

Half an hour's sharp trotting brought us to the beginning of the long avenue which led to our entertainer's noble mansion. On arriving we had a hearty Highland welcome from him and his assembled guests; but the hour being already somewhat late, the necessary introductions were hurried over, and then we were soon following the head keeper down a winding path into the valley below the house.

The morning was lovely, cold, and clear as could be wished, while our "fighting line," winding through a deep forest of firs, was really a picturesque sight. First went the keeper in his national dress, a man of strength and stature, and an awe to all the poachers far or near; then our host, P——, discussing the merits of a new trout fly with an Assam tea planter, R——, whose gun, carried over his shoulder, had recently been dealing out death and destruction to snipe on the plains of Northern India. On their heels came our host's son talking to "Uncle P.," as he called that relative of his, and two cousins, both in Athole tartans. These, myself, J——, and one other young laird made up the party. We wound down the narrow path in single file, the occasional gleams of sunshine breaking into the cool shade of the forest to glitter on our gun barrels. We chatted and laughed until, having dipped into a lovely glen, thick with amber fern and silver birches, we crossed a rocky torrent bed, scaled the opposite bank, and soon found ourselves by a thatched cottage, where keepers with numerous dogs in lash awaited our arrival.

Now chaff and fun had to be given up, for we were about to begin the serious business of the day, and our host, an unwavering enthusiast, led us out of the wood, across a patch of rocky ground, through a gap in a stone wall, and there we were on the breezy hillside, knee deep in heather, breathing such nectar as dwellers in towns never dream of, with in front a limitless expanse of mountain and moorland undisturbed as far as the eye could see by a trace of civilization. "Can this mighty, uninhabited expanse be in the overcrowded British Isles?" I wondered; but my host "sniffed the scent of battle afar off," and stopped all musing by an imperative "Come along!"

Our first position was behind a broken-down stone wall, where the keeper dropped us some seventy-five yards apart, and with our faces all to the eastward whence the birds were to be driven up. This turned out to be but a poor sort of cover, for though the wall in front of each shooter had been built up to serve him the better, yet to be out of sight it was necessary to sit or crouch down, either of which positions are fatal to good, rapid shooting. The best screen in driving game is always found to be one that comes up to the neck of the shooter when standing, thus allowing him to turn rapidly and give him a clear shot in every direction. We occupied our "marks," such as they were, and making ourselves comfortable awaited in silence the arrival of the first bird, amusing ourselves meanwhile with our delightful surroundings—numberless mountains fringing in an amphitheatre of purple moor, all rugged and grand, some just tipped with snow at the highest points, and gleaming silver where the sun lay upon them, and purple in the shadows of the ravines. The wind from these snow-fields, now that we had no trees to shelter us, was as cool and fresh as it could be, sweeping over the wide expanse of moors, and bringing to our ears the far away bleat of mountain sheep, or the melancholy whistle of a plover, whose sharp eyes already perceived the advancing beaters. But the sun was warm overhead, and our pipes smoked fragrantly, so we waited with contentment for the battle to commence. Presently a distant shout comes down to us, and the guns all down the line are to be seen directly on the *qui vive;* cartridges are hastily arranged, caps securely "crammed" down on their wearers' heads, and all eyes are directed over the wall to get a wider view of the plain in front; and soon the grouse come in sight on the far left of the line, giving the last man one chance, and his gun immediately breaks the silence of the hills, the white puff of smoke sailing away over the heather to leeward. Then some blackgame go over to the right under a regular fusilade from the batteries down there, and it becomes obvious that though we cannot see them, yet the beaters are all among the birds down the hill slope.

Soon my turn comes, and I see R—— making signs to me under cover of his ambush and taking a peep at the moor in front; there is a large covey coming "dead" for my stand. It is always an exciting moment, even to those who think little of driving as a legitimate sport. The birds appeared skimming lightly over the tops of the heather, seeming almost stationary for some time though travelling at a great pace, and little is to be seen of them but the head and narrow edges of the outstretched wings. Another second or two and they are within forty yards, and as my gun speaks the foremost bird drops, the others going at such a pace as on such near acquaintance as we are now seems terrific, rise to clear the wall, passing overhead like meteors, in another second are retreating over the heather behind the line. I fired again, R—— fired, my brother fired, his bird coming down within a few feet of the stand occupied by me; and to our astonishment, when we thought it was all over, "Uncle P.," far away down the line, also sent a couple of charges of shot up in our direction, but without bagging either men or grouse.

We get a few more shots, and then the beaters arrive, the retrievers are unslipped, the slain picked up, after which we walk in line over some rough ground, where the dogs find another bird or two and put up a lowland hare which our host stops in workmanly style.

At the next broken-down dyke we disperse again to our posts, spending the interval, while the beaters walk round the moor, in adding to the screens as our fancy suggests, and making our seats comfortable in the manner set by our luxurious friend the Assam planter, whose first care at every stand is a springy nest of heather, on which he reclines in bliss until the birds arrive. Again the same sort of process is gone through, and a rather long wait well rewarded by a rush of grouse, mixed with small bodies of blackgame, hares, and squadrons of shrieking plovers, when the beaters get within feel of the enemy.

The cannonading is soon brisk up and down the line—the two young gentlemen in tartans getting a little "off their heads" with excitement, and showing themselves freely (a great mistake in grouse driving), sweep the neighbourhood with their well-served guns, while "Uncle P.," who, by a judicious and philanthropical foresight of the head keeper, is always their companion, far away down on the left, also gets a "wee bit daft," burning much powder with great satisfaction to himself but little effect on the bag. We up in the centre, however, behave ourselves with decorum, never firing at any birds but our own, and carefully making a mental note of where such of them as we may bring down will be found when the beaters come up. I have heard of this latter matter being settled in a very cut-and-dried manner with the help of a pencil and sheet of cardboard, the latter being divided by lines into quarters, with a circle where the divisions meet in the centre to represent the stand; the shooter carries a supply about with him, and, dividing his neighbourhood at every drive into imaginary portions, marks with the pencil as nearly as he can the vicinity of every bird, as he brings it down, on the sheet of paper—a cross for dead birds and a dot for probable runners, this record being handed over to the keepers when they come up; an arrangement, I fear, which though it may read well enough, would need a shooter as many-minded as Cæsar to carry out in the heat of the fight. By the time the sun high up in the sky points to a little past midday, being all more than ready for lunch, we seek a sheltered nook, cut deep through the moor by the ceaseless labours of a sparkling streamlet, where, on a broad, sunny rock well out of the wind, we find luncheon spread and our host's charming daughter in the neatest and most reasonable of costumes ready to welcome us, while the big mastiff at her side makes hill and valley echo to his sonorous baying until a sign from his mistress's hand informs him we are lawful intruders, when he forthwith subsides into the heather.

It is by no means the worst part of the day; the provender is ample and varied, cold grouse pies, flanked by such salads as must surely have grown in celestial kitchen gardens, a sirloin of the finest stalled beef, pastry of fairy lightness, unimpeachable drink that, when accepted in foaming tankards from the fair fingers of our fascinating Hebe, becomes quite ambrosial. We linger, too, over the choice cheroots which our host passes round after the meal; thus careless of time until the edges of the purple shadows creeping up the opposite hillside warn us that autumn days are all too short for much idleness, so we see the "mem sahib" to her pony carriage in the neighbouring lane and then are soon hard at work once more.

The first wait afternoon is a long one, the keepers and beaters seeming to have lunched as well as we have and to be rather lazy; however, we are contented and sit calmly in our shelters, our guns across our knees and the position of each man down the long line of grey wall marked by a tiny curl of tobacco smoke ascending in the still air, for the morning breeze had died out as it often does in the latter part of a Highland day, and all the wide, lovely landscape before us simmering in the golden glow of the quickly sinking sun.

But after twenty minutes or so there comes a shout mellowed by distance echoing over the corrie, and soon a devoted band of little brown birds are on the wing coming along all in a bunch. They come nearer, and are just within long range, the cock bird leading and the rest "twinkling" over the heather behind him, when the report of the gun of some impetuous individual, whom we have no time to see, disturbs the stillness, and as the covey breaks up to right and left we all get our chances, thinning their numbers until they are out of shot behind us.

Other drives follow, bringing up the bag to a very respectable total, considering the lateness of the season, but so much alike in the details of the slaughter of the unsuspicious little brown birds "butchered to make a Roman holiday," that it would be but tedious to narrate them all; and then we have finished the final beat and troop homeward as the sun sets, not quite so noisy as in the morning, but well pleased with the day's shooting. Nor are our consciences, whatever the tender-hearted may suppose, overburdened with the manner of our sport, for we feel that at this time of year we could not have got near the birds in any other way; and finally, as our host remarks with a sigh, handing his gun to the keeper, "It is the last bustling they will get until next August."

CHAPTER VI.

# PARTRIDGES AND PHEASANTS.

It is a mistake to think Hodge hates game and game preserving! Such hatred is done for him by those political hot gospellers whose forum is the poacher's tap and larceny their chief creed. The agricultural labourer himself rarely comes within touch of the birds whose names head this chapter. They do his allotment plot no harm, they excite none of that savage envy which the village charlatan would fain propagate in rising under his feet on stubble or gorse. Occasionally he gets a day's pay for a light day's work as a beater to some shooting party, but otherwise his interest is all metaphysical and remote. When his hot gospeller mounts the rostrum, Hodge listens open-mouthed to the new science of spoliation, and in his inner heart wonders so neat a gentleman can lie so eloquently as it is palpable his tutor does.

The average countryman knows, in fact, it would not swell his score at the village post-office, to wipe out game from the land. His interest is linked with that of the farmer, and the latter, as his spokesmen freely acknowledged before the Game Law Committee, is advantaged greatly by the popularity with the money-spending classes, which game brings to our counties.

The rustic could never look upon game as a food supply, —our shires would not stand his demand for three years if coverts and stubble were thrown open;—the pleasant freedom of trespass and opportunities of collecting faggots of other people's wood, or gathering mushrooms or hazel-nuts, are trivial gains to compare with what he would lose.

As for the farmer, what could he gain by doing away with shooting. Ground game is in his own hands (and little good it has done him!) and for every peck of wheat pheasants take along the wood-sides it is the fashion now to over-compensate the grower. Is it possible he does not generally appreciate those magic dates, September 1st, August 12th (if he is a dales' man), and October 1st, and perceive the economic value and boon of a fashion and a health-giving pastime which calls back peer and commoner from Norwegian trout streams, from the soft seductions of yachting in southern seas, and from every quarter of the globe to consume his beef and mutton, to buy horses, and to send up the price of his seed and grass! In Ireland the want of resident landlords (they shoot the few they have) is half the trouble of the land, and in England matters will be even worse when the gentry have been brought to spend half the year in town and the other half at Monte Carlo or the Riviera.

Sydney Smith often thundered against the game laws in the press, and that amiable bigot, George Grote, thought he might rise to notoriety in the parliament of our grandfathers by adopting a like line. Since then this bone of contention has been well wittled; yet there are some pariahs in politics still intent on it!

As far as I can see, their aim and object is to do away with pheasants and partridges, for, infatuated as they are, I can hardly think them so misguided as to suppose these birds would remain amongst our fauna half a dozen years as *fera naturæ*. And were the obnoxious laws abolished to-morrow, sterner enactments against trespass would be essential, as pointed out in Mill's "Essay on Government." As an example of the aims of this new school of thought, it may be mentioned that under the Ground Game Act of

1880, the occupier of moorlands and unenclosed lands can only kill ground game between December 11th and March 31st. Yet these poaching politicians propose to enact that "the tenant of a hill farm" may shoot, or authorize his sons, friends, and, if he chooses, every poacher in the neighbourhood to shoot, not only ground game but also everything that flies, on every week day throughout the year. A correspondent points out that the tenant in question will be able, if Mr. Menzies has his way, to sally forth, accompanied by his myrmidons, and by a lot of curs with keen noses, but under no control. A hare, or perhaps a bird of any kind, springs up. Away go the yelping pack in pursuit. "Conceive," says the correspondent, "this going on almost daily when the grouse are on their nests, or when the young are just out of the shell, and continuing steadily until the 12th of August, and so on throughout the season." The tenant is also to be permitted to shoot grouse and blackgame upon his stubbles. Who can doubt that in many cases, tenants will spread "stooks" of possibly worthless grain to attract birds from the neighbouring moors?

Again, though perhaps a little foreign to the subject, I must quote the sensible and pointed words of a public writer who, in view of this radical propaganda, says, "Recent measures proposed for Scotland are mild, however, compared with that which some would enforce in this hapless country. In the 'Ground Game Act (1880) Amendment Bill,' proposed by the four English members, 'any owner endeavouring to induce an occupier to forbear to exercise the right to kill ground game is liable to a penalty not exceeding one hundred pounds nor less than twenty pounds.' By this clause all contracts about game between a landlord and his tenants are rendered illegal. By another clause, 'No person shall kill or take a hare between March 1st and June 1st in any year.' This provision chiefly affects spring coursing, and we must leave the managers of Kempton Park, Plympton, and Gosforth Park meetings to digest it as they best may.

Without following these proposed measures further into detail, and pointing out that their inevitable effect will be to clog game preservation in England and Scotland with such restrictions as ultimately to lead to its total abolition, we would fain address one question to these, doubtless, well-meaning Don Quixotes. Do they suppose that the British isles contain all the ground available for game which exists upon earth? In the face of such legislation as has lately been passed, and, still more, of that which seems to be in contemplation, what inducement can there be for men of wealth to retain their existing properties, or to acquire new estates in this country? 'The world is all before them where to choose their place of rest;' and either upon the North American or the Australian Continent, or even in many parts of Europe, they can easily pick up vast domains 'for a song,' where game of all kinds swarms, where the climate is preferable to our own, where taxes are low, and where Puritan legislation will never disturb them. Eliminate from England and Scotland their resident country gentry, and what will there be left?"

But though I feel strongly that game ought no more to be done away with than soles and flat fish round our coasts, or a fancier's rabbits in his back yard, yet no one can recognize more keenly than that if pheasants have a right to live, so have peasants. Because I feel strongly the Charybdis of game destruction, as known in Switzerland, is impolitic and foolish, yet on the other hand, the Scylla of preservation, as illustrated in Persian game laws, is equally unpleasant. The happy mean is what must be aimed at in such matters, and this can only be discovered by reasonable and neighbourly discussion.

I have felt the strongest indignation at meeting notice-boards round highland glens declaring the *free* heather the private privilege of an unappreciative landlord, and have sent to perdition more than once the "owners" of delightful trout streams who have pretended bastard rights to close them against every other of her Majesty's lieges. I can quite understand, moreover, that in the south a sylvan Eden behind a well-spiked, six-foot oak fence can and does stir the radical gall and germinate a hatred of those costly "wild fowl" for whom the demise is kept so quiet. Yet these gentlemen should know general confiscation is a poor corner-stone for the erection of a temple to freedom. Were reflection in their line, I would refer them to the epitome of foreign game laws at the end of this volume, wherein it will be seen that in all countries, even those happy, blameless, and Arcadian republics of France and Switzerland, game is preserved with more or less rigour.

But reason is not in their line, so perhaps the best thing we can do is to educate the misguided rustic, or the thick-headed voter who has listened to these too reckless ones, until with Queen Titania, they exclaim—

> "Ha! what madness has possessed me!
> I dreamt I was enamoured of an ass!"

Meanwhile the sportsman's birds are being more scientifically reared, and more carefully tended year by year. There is no perceptible thinness in the rows of pheasants or other game which fill our poulterer's shops during the season, and if prices remain high, it only indicates the constancy of the demand.

Some remarkable bags were made during 1885. At Elveden, the Maharajah Dhuleep Singh's, on the borders of Norfolk and Suffolk, 3258 brace of partridges in fifteen days. On Lord Walsingham's estate of Merton, bags averaging 200 brace a day were made at the beginning of the season. On this same estate 2000 *wild-bred* pheasants have been killed in a season, and this after innumerable nests in the open had been despoiled of their eggs. In fact, Norfolk and Suffolk, with their great game estates of Euston, Merton, Riddlesworth, Buckingham, and Wretham, are the best and most prolific game counties in the world. The

Maharajah's estate is, as I write, in the market; and Dhuleep Singh himself told me, in course of conversation before he left for abroad, how much he felt the parting with this admirably managed domain where he has shown the *ultima thulæ* of scientific rearing, and beaten all records in the number of game brought to bag in a day or in a season. When I laughingly asked him if he was going to cultivate the chickor or preserve sand grouse in Runjeet Singh's fertile plains, he shook his head despondingly, and suggested the near future might see a sterner sport on the Indian frontier.

Here at home there is no break in the value or sportsmanly estimation of shooting and shootings. As far as the lesser species is concerned, if they escape politicians, I cannot see why they should not multiply and flourish greatly. Our last agricultural returns show that the kingdom is slowly but surely turning to a grass and orchard country, and glebe and meadow in maugre of a few sceptics is by no means adverse to the russet birds. Thus one sportsman writes—"I must take exception to the idea that a grass country, no matter how well preserved, rarely affords good partridge shooting. Some of the best partridge shooting I had some time ago was over a large tract of grazing land. I found it well stocked with birds, and, being under an impression that the grain of a tillage farm was necessary to their subsistence, I opened the crops of the birds, which were very fat, and found nothing but grass seeds in them. The long grass and ditches and briar-covered dykes afford them plenty of shelter."

I can endorse this, as probably many other sportsmen can, having brought down partridges as thick and heavy on the grassy Hampshire downs as any that ever came from Essex flats or Suffolk turnip-fields. Not only is wheat not absolutely necessary for their constitution, but where cereals will ripen in the Highlands partridges may be established; and any laird who would try should get a few sittings of eggs next spring, hatch them out, and rear them with bantams, taking care to keep the broods well apart as they grow up, to prevent them from packing, for if they do so they often, when disturbed, take uncommon long flights, and on that account they may leave his grounds altogether. Where cultivation at a considerable altitude in former times has been going on, but where there are now only a few green patches of grass land, the rest all heather and bracken—a few broods of partridges are often still to be found. They are usually smaller in size than their more favoured brethren in the low grounds, and their plumage is considerably darker; their flesh is as high-flavoured almost as grouse to the palate, no doubt from the food they subsist upon. I have shot them in winter with their crops full of heather.

I do not believe, with Mr. Greener, that there is more than one species of English partridge in England; with a bird varying so much in location and food, great varieties of plumage and build may be expected. The French species, one of our most beautifully plumaged birds, and weighing sometimes as much as 1 lb. 6 oz., here and there proves an enemy to the home birds, much to the resentment of shooting lessees and the spoiling of dogs, as the bird is an obdurate runner, and will, unless headed, slip from field to field without rising to "tempt the hazard of the die."

The naturalist sympathizes, and sees no more reason why a "red leg" should not run under such circumstances than an Irish landlord in like case; but then the friendship of the naturalist is indiscriminate, and as he cannot explain the diffusion of species he would like to see them still more diffused. Thus he is much in favour of that pleasant subject, acclimatization, which, however, is too wide and curious to be seriously entered upon here. A few erratic attempts, it is true, to enrich our home fauna with game birds likely to vary our shootings without enlisting the hostility of farmers have been made, but not, we think, with much judgment. Examples of what might be done are obvious.

The brown partridge is no doubt indigenous, but the pheasant certainly is not. Echard thinks it was brought into the kingdom during the reign of Edward I., though perhaps much earlier. Then if quail have been with us since British times, the French partridge is new, and though ptarmigan and grouse are as old as their strongholds the hills, the noble capercailzie has been successfully restored to Scotch forest from which sixty years ago he had died out.

When Ireland returns again within the confines of civilization, her willow and spruce wastes certainly ought to be stocked with blackgame, which will flourish on ground where grouse would starve. In the south of England we have hundreds of thousands of acres of barren heath and bog myrtle, upon which the few blackgame, formerly to be found, are nearly extinct. We are most anxious to see some bird worth powder and shot occupying these wastes. "I have been long anxious to see the introduction attempted of the Scandinavian species," says a correspondent. "From what I hear, not having visited Norway myself, I believe that with a little co-operation there might be a fair chance of acclimatizing these species of grouse in Hants and Dorset if attempted simultaneously on the crown lands and by some of the chief landed proprietors." Knowing Norway myself fairly well, I should doubt if any species, accustomed there to its luxuriant pastures and great feeding ranges, would settle down in necessarily circumscribed English barrens, though the experiment might conveniently be tried.

American prairie grouse for our wild pastures have been much talked of without practical result. Mr. G. H. Bates, in the *Field*, seems to be a strong partisan of this species.

"PRAIRIE CHICKENS FOR ENGLAND.

"SIR,

"Yesterday a friend and myself trapped, alive, twenty-one prairie chickens. I have just eaten a hearty dinner of the same, and while doing so it occurred to me that 'John Bull' could, if he chose, soon have of his own raising hundreds of thousands of these useful birds to shoot at and to eat. There are thousands of square miles in Great Britain where these birds would do well; in fact, I believe that they would multiply much faster either in England, Ireland, or Scotland than they do here. They are very hardy, and not at all destructive to field crops. The hens commence laying about the middle of April, hatching in June. They produce at each sitting from twelve to thirty young; I believe I have seen a greater number than this in one covey. The average weight of prairie chicken is about 5 lb., with a slight increase in the male. From now until April they can be secured alive in traps in great numbers, and I believe that they can be delivered alive and in good condition in any part of Great Britain at a cost not exceeding 10*s*. each. Then, why not have these 'Yankee chickens' of the West, in countless numbers on the downs, on the moorlands, and in the evergreen forests of Merry Old England? If twenty or more gentlemen, owning estates in different parts of England, and an equal number owning estates in Ireland and Scotland, would subscribe for two thousand or three thousand of these chickens, to be divided equally among them, they would confer a great benefit on the people of Great Britain, for their action would in time add an important element of supply to the tables of both rich and poor. Two thousand or more birds can be caged and sent in one lot, and, by having a proper person in charge of them, very few of them would die during the passage from the West to England.

"Import two thousand live and healthy prairie chickens into England, Ireland, and Scotland; let them loose in grounds favourable to their existence, and they will produce more of their kind in seven years than there are at present inhabitants of Great Britain. Should any person reading this desire any information that I can give, or should any person desire to experiment by importing prairie chickens into England, I will do all in my power to aid him.

"(Sergeant) G. H. Bates."

"**Saybrook, Maclean County, Illinois, U.S.A., Dec. 13, 1883.**"

But some of our own authorities decry the bird as of ignoble, skulking habits. Mr. Harry Greenwood says one could hardly wish for more delightful or better shooting than the Virginian quail affords. He flies like a rocket, and in the superb autumn days of America, when he is full grown and strong, and vigorous on the wing, it requires both a true eye and a rapid hand to cut him down.

Other birds have been recommended, nor can there be any reasonable doubt that if the opportunities were forthcoming we might find amongst sand grouse, the lesser game of the Mediterranean shores, or even the pheasant-haunted rhododendron thickets of the Himalaya, some birds that would thrive and multiply amongst our coppices and downs. Meantime we have enough game at hand to fulfil every reasonable need. The farmers, who must ever have much of the sportsman's enjoyment within their control, are for the most part wisely content to let things remain as they stand; and if the veto of ignorant politicians receives the contempt it deserves, we may still for many years hear the crow of the cock pheasant as he comes out to feed "in the dewing," and notice how tenderly the partridge cherishes his lavender and cinnamon bride, how faithful and gallant he is to her and to his school of dainty striped chicks who people the corn and clover lands, and nestle under that "field of the cloth of gold" which means the yellow harvest of fertile England is ripe once more.

## October by Covert and Hedges.

From heather to stubble, and then from corn lands to oak coppices, and, later on, the holly thickets for woodcock, willow-fringed brooks and reedy margined estuaries for ducks and winter wild fowl, is about the sequence in which hillside and cover alternately interest the sportsman. In autumn he has done with Perth and Rosshire, or he has very nearly done with them. There is a thin skin of snow, perhaps, on the caps of the highest peaks around his moorland lodge, and a suggestive hoar frost or two has touched the bells of the heather, making the bracken undergrowth of the birch and spruce thickets, beloved of the rabbits, glow in amber and red brown.

Whipping for trout he finds, in an open boat, with a gillie to row and take all the warmth-giving exercise, has become but a chilly pleasure; so the trout have a holiday, at least in the very far north, while grouse and blackgame can exchange their opinions upon the last shooting season without fear of much further interruption.

As for the stubbles, some of the most enjoyable sport on them is yet to come in favoured localities. We know the midland turnips have been assiduously stumped by the orthodox squadron of shooters in line with their beaters and dogs, and Suffolk has driven its broods hither and thither over the hedges, rich in their harvest of scarlet berries, behind which the latter-day sportsmen are content to stand and enfilade the coveys as they rush by; but in spite of all this, there is plenty of pretty shooting still left on the frosty mornings of early winter, when the sun gets up behind the bare ash thickets, a heavy red ball of fire, the stubble is brittle as glass underfoot, and the fieldfares are quarrelling noisily over the ivy berries and haws. Then, with a clever setter, as the day warms, we may try the woodsides and the low-lying rushy meadows for partridges with every prospect of success.

But October and November are properly the pheasant shooter's months. Theoretically, no doubt, the season begins on October 1st, but the truth is the mild autumnal rains with which September makes fresh and tillable the summer-scorched ground, give a new lease of life to vegetation, and thus the woodlands are often as thick when the close time ends as they were in "leafy June." Generally it was so, and, as a consequence, though we may go round the outlying gorses, the main body of birds find security and a stronghold in the coverts, until the verge of November. By this time, if the game-room is not full it ought to be. Such, at least, was the opinion of a friend of ours the other day when he wrote and begged we would come over for a preliminary "dusting" of the pheasant, and lend our assistance in the replenishing of his larder, upon which hospitality and generosity made constant calls; and, though we knew the sort of shooting—to-morrow would not be our idea of perfect sport—we went.

Reasonably early next morning, our host, L——, J——, and myself, are seated at a substantial breakfast in the dining hall of the house, laying in a foundation for the day's work. But when the meal is over, and we have betaken ourselves to the smoking-room for a quiet pipe, we stand, hands in pockets, looking out of the windows, while our courage sinks at sight of a steady November downpour almost hiding the landscape. Yet the game room requires replenishing for an approaching festival. As J—— remarks, it is not the rain we mind so much, but will the birds rise or the beaters work properly in such weather? However, presently our host throws away the stump of his cigar with an impatient ejaculation, asking whether we are to stand watching all day for a bit of blue sky, which obviously is not coming, or if we will brave the elements. We vote for a move at once, and while donning our waterproofs, the shooting trap comes round to the door with a pair of strong bays in the traces. L—— takes the reins, I mount beside him, while J—— and the footman scramble in behind, with the guns under their feet.

Fortune is a fickle jade, it has been somewhat disrespectfully observed, and no sooner are we off than a glint of brightness comes over the sky—the clouds tear apart in the southward, and soon sun and shower are coursing alternately over the meadows, beautiful in their mellow browns and greens. Away we rattle through the park, deerstalking caps drawn down, and collars turned up to our ears, the horses fresh and frolicsome, dancing and tossing their heads, their bits rattling, and bright gold trappings jingling gaily as we fly down the smooth gravel roadway; through the "chase" towards the first of the woods to be beaten. It is not far, and in twenty minutes we turn up a shady lane, ducking our heads occasionally to avoid the acorn-loaded boughs of the thick hedge-row oaks, and arrive on the outskirts of a wide tract of undulating woodland, broken into by patches of cultivated ground. We take our guns, order the luncheon cart to meet us at midday, and join the party of beaters coming down a "drive" in tail of the head keeper. The latter is despondent but alert. He touches his hat in obvious pleasure to see us out, but can give us only small hope of sport. It is raining again now, and we have the additional discomfort of drip from the trees; but we pull down our caps, and securing every button of our waterproofs, determine to face our luck. The first stands are in a disused and moss-grown roadway, the beaters swinging round and beating back to us through a long strip of gorse, fern, and scattered beech bushes. Tap, tap, go the sticks on the wet shrubs, doubtless bringing down upon the luckless beaters, showers of moisture at every blow, while we, hardly better off, keep running a finger along the midribs of our guns, to free them from big drops which continually accumulate there. Little stirs for a time save a blackbird or two, and we stamp about somewhat impatiently, for we are cold and benumbed about the hands; yet we know there are pheasants and rabbits afoot, for we hear an occasional shout, with renewed tree-tapping, as the men keep the quarry from breaking back. Slowly the game is driven down to the end of the strip, where we know there is work for us. A couple of thrushes bustle overhead, fluttering J——'s nerves considerably. Then the usual rabbit, at its wits' end to account for the hubbub in these well-guarded woods, bolts across the roadway without the challenge of a shot. Two pheasants followed, rising at once almost perpendicular from the yellow ferns into the air, and attempting to clear the hedgerow hazels, under which the game is now closely hemmed in. Both birds fall, and now the work for some time becomes after its kind exciting. The pheasants, running about hither and thither amongst bramble tangles, or wild rose thickets, put off the much-dreaded flight until they can no longer find any shelter. Then they get up with a swiftness and decision which perhaps redeems the sport from the mere slaughter it would be otherwise, by making it by no means easy to stop them properly before they are behind the oak trees. We are fairly successful, considering the difficulties of weather and position, and soon there are half a score of "gorgeous slain," with two or three rabbits, to be picked up and handed over to the care of the under keeper as a result of a first drive. The process is repeated in another direction, when the guns take places along the edge of some "springs" or sapling oaks, waiting with as much patience as may be, until a blackbird vidette or two, flitting hurriedly overhead, tells that the beaters are approaching. Then the usual excitement and "fusilade" comes off, with more or less satisfactory results.

During these autumn weeks nearly every bit of woodland in the country sees something of this sort. While unquestionably not the highest development of woodcraft, properly conducted pheasant shooting is by no means easy and certainly not cruel sport. It is almost the only time when the shooter finds his recreation in woods, which, before winter has completely stripped the trees of their golden spangles, are so proverbially lovely. The surroundings of his amusement are delightful, though probably close to home; his victims are noble birds, and strong winged, even if they are hand reared, nor by any means so easy to "stop" as is often supposed. But all long tails do not come by their end in the rain, or under the heavy handicapping of compulsory flight over well-posted gunners! There is that stray pheasant we come across in the hedgerows when we have been out shooting something else, who beguiles us into a long hunt down the fallows until he gets into an old gravel pit in company with a couple of hens, and a hare perhaps, and gives us a very pretty shot as he leaves it. There is the wandering bird who gets up under our noses far out in the open marsh lands, from a spot far more promising for teal or snipe than for any of his feather. We have even put up pheasants when shooting grouse on heather, where never a bush, much less a spinney or plantation, was within sight for miles around. There is, indeed, hardly a place into which these birds will not stray, and the rearer of game knows this, and he must take the precaution to feed them well at home if he is to keep within his own boundaries those birds which will cost him little less than half a guinea a head between egg-laying and larder. Personally, as we have said, we do not much care for "corners" at cover side this autumn weather, be it bright or misty. Birds at twenty shillings a brace, though they cost us no more than their powder and shot, are too much for us. Rather we prefer the fair quarry of a strong-winged Exmoor cock pheasant, who keeps his look-out amongst the stones and fern of the tors and turf hills, and gets up with the noise and vigour of a bird of four times his size. And we like those sea-shore pheasants of the Devonshire combs, second to none in beauty of plumage and robustness, which haunt the undercliff, and feed down to high-water mark. We have had as pleasant a ramble as could be desired, again, after the cocks that come with the snow and frost, from goodness only knows where, to the rhododendrons and yew hedges of Scot manses, or the laird's outlying stackyards and last year's lambing pens. On all such occasions of wandering sport the resulting "bag" would look unconscionably foolish by the side of even a poor day's work outside a midland autumn wood but recently populated from Leadenhall Market or game-rearing Suffolk! Yet we fancy, in this case at least, there is little wisdom in numbers. Everything tends to show that the Englishman of the next ten years must be moderate in his views of sport, and for ourselves we are ready for the change. Nothing, we think, endangers at present our legitimate shootings in woodland and stubble so much as over preserving, and debasing an old-fashioned moderation in the desire to fill up our game books and shoot a few more head of birds than our neighbours.

Perhaps it would be too serious a retrogression to shoot pheasants as Lord Byron did at Six Mile Bottom, to revive the slow hunter pointers with bells round their necks of one writer, or even Lord Middleton's "pottering clumbers;" yet there was good relish in our forefathers' sports!

For them their woods did not mean "one crowded hour of glorious life"—and then idleness. The bloom might be off the shooting season; the renter of moors had had his last "drive," duly thinned out the weary old cock grouse, closed his lodge, and come southward; the keenness of the partridge shooter for his special game was somewhat dulled, yet there was still shooting to be had in the shires, and very pretty work for the moderate-minded devotee to the gun.

Harvest over, of course, the all-involving steel of the reaper had long since ceased to shine amongst the miniature forests of tall yellow stalks, sweeping to one common ruin the trembling crop, the gaudy blossoms of the poppies, the delicate sisterhoods of the climbing convolvulus, and a world of such tender vegetation that put its trust in the shelter of the giant grain around it. Yet with harvest commenced the old fashioned lowland shooter's campaign.

Who is there who amongst his earliest sporting achievements does not remember the delights of his first bout with the rabbits in the stubble fields? He must recall with enthusiasm those outlying rabbits in Farmer Wurzel's fifty acre plot, who had stolen at earliest dawn into the patch of corn left uncut from yesterday's work, meeting their fate from the muzzle-loaders of the farmer's boys before the sun was clear of the elm tops. They thought, perhaps, the square of wheat was cover enough when the first of the workers came filing up through the lane into the field at six a.m., bolted into it incontinently, and had hardly a misgiving or a guess at how serious affairs were becoming for them until the whish of the scythes grew closer and closer on every side, and yellow daylight came down the furrows that had lain before cool and damp in the green gloom of flowering herbage. Then they made the delayed but unavoidable rush. Even such humble sport—the carnival of the half-shorn corn—was good fun.

Bunny number one was "chopped" by an active lurcher called up from guarding his master's lunch under the hedge to take a share in the sport. Number two might have reached the fern clump he was running for had fortune been kinder. But, alike to heroes contending on Trojan plains and conies delivered over to the sportsman, fate is sometimes cold and forgetful of the brave, and thus he rolled over to a charge of "6" to be soon stretched out by his comrade. Number three *did* get away, because he had the sagacity or good luck to dash close past the worthy farmer's well-legginged legs, and his boy, Master Wurzel, was too dutiful to risk the chance of "peppering the guvnor," though the provocation was great. So the tale went on of the rough game; some rabbits meeting their fate from the guns posted at the fast drawing-in corners of the square, and others coming by it in less legitimate fashion.

After this dusting of the rabbits comes the feast of St. Partridge, with its hot tramps over bare stubbles in search of "wee brown birds;" and then the pheasant shooter's chance, the competition for good corners in the pleasant amber-tinted woods, the tipping of authoritative velveteens, and the "heaps of gorgeous slain," as the daily papers have it, laid out in the game rooms of country mansions. And finally, when only the outermost twigs of elms are tagged with yellow leaves—well—there is still something to be done.

We will suppose a shooter rises moderately early—for enthusiasm must be tempered by experience in this matter—while the October mist lingers in quiet hollows, and gossamer laces the tall grasses and bramble sprays. He breakfasts, and forthwith sets out, if of rational and quiet inclinations, with a well-chosen friend, a brace, or perhaps two, of steady setters, and a couple of bearers. Such an array combines the comfort of individual sport with reasonable chances of securing a good bag. Large parties I abhor, while, on the other hand, two or three guns, with bearers between them, are the least which can be reasonably expected to give a decent account of the game flushed, and properly work the ground in an open country. So we will set to work, five of us, all told, including the keeper. We have hardly swung our legs over the gate that separates our first field from the road, when up spring a pair of partridges from the cart ruts in front, twisting themselves over the twelve-foot hawthorns before we can draw a head on them. A little *contretemps* of this sort is by no means without its useful purpose. The guns forthwith pull themselves together, keep their weather eye open, and if again taken unawares fully deserve all those hard things that will be said of them. A little further on, out of sight of the footway, we bag our first birds, an old cock, who sat overlong, wondering whether we were harvesters come back again or open foemen, and two youngsters who relied in the parental sagacity. We are not above picking these birds up and stroking down their exquisitely blended plumage of grey and russet, provided our proceedings do not unduly interfere with the comfort and decorum of the line. We greatly resent having our game snapped up and crammed into that leather atrocity, a hot game bag, before we have given a glance at it. Surely if a man neither works his own dogs nor sees the birds his gun has brought down, he might be just as profitably employed "blazing" into a bandbox full of sparrows. By far the happier method of business is to take a personal interest and a personal responsibility in all going on. These English partridges feed chiefly during the quiet hours after sunrise and before sunset. When, in the morning, the head of the covey has stuffed out his russet gorget with a sufficiency of ripe seeds and a flavouring of insect life perhaps, he betakes himself, with all the members of his household, to a dry and sunny spot where digestion and meditation may go on undisturbed. Birds will be found in the fallows and green crops at all hours of the day when they have been disturbed. A little practice soon enables the keen sportsman to know by instinct where they ought to be, and it is curious how seldom his intuition fails him.

We beat two or three fields with fair success, and try them along the warm side of a "hanger" or sloping wood, there making some varied additions to the bag. The rabbits, for instance, have come out, and lie close until disturbed by the dogs, when they flash across the green turf of the roadway under fire of the innermost gun. Wood pigeons, too, undertake rash peregrinations from amongst the last crisp yellow leaves of the Spanish chestnuts, and are tumbled unhesitatingly amid a cloud of feathers into the fern and ling. Surely there is no such bird as the ringdove for shedding its plumage. We have shot many, and yet never one that had its feathers fixed to its cuticle with any reasonable firmness. A hare occasionally comes deliberately out of the ditch and limps along till a stop is put to her vagaries, and the keeper parting the strong sinews of one leg, thrusts the other through the opening, carrying the big beast thus on a stick; she is too heavy and gross for delicately feathered company. A pheasant is, mayhap, the next bird that falls, rising behind the leftmost gunner from a clump of oak springs, never struggling after he was struck, but coming down heavily through the bare saplings.

These coppices of springs or knee-high oak are capital cover for game, especially when, as is often the case, woods adjoin. The ground is uneven and broken, with the commencement of gravel-pits hardly deep enough to hide a standing man, and holes whence stumps have been extracted. Round the mossy rims of these are the hares' beaten footpaths, and where gravel and fine sand collect under overhanging brambles, partridges come in from pastures and fallows of a morning to dust, being, *sub rosa*, often trapped there by the knowing village poacher. This region boasts many faggot stacks, strongholds of the weasel and stoat, and building places dearly beloved by the industrious wren. Its soil is deep vegetable mould, covered with an accumulation of leaves and dried twigs, all to be added to the peaty store in future years. In spring, bluebells carpet it as far as the eye can see, making another firmament for the pale starlike anemones and primroses to shine in. A touch of admiration may be permitted to one who knows and loves such spots so well. It is in these glades the nightingales muster strongly each spring. We never listen to them without thinking of Isaac Walton's tunefully expressed delight in their song, for love-song it is. "Lord!" he says, "if Thou allowest such melody to bad men on earth, what music hast Thou prepared for Thy saints in heaven."

At this time of the year the earth underfoot is perhaps a trifle soft, but our boots are stout, and little we care for that. The rabbits sit in every tussock of yellow fern, only bolting out on the most pressing invitation. They dodge amongst the brambles, a vision of grey fur and white "scut;" two bounds take them across the deep ruts of the woodland timber road, and utilizing the trunks of those oaks whose girdles of blue or red paint, like the cross on the lintels of the Israelites, has saved them from destruction, they use every effort to get away into the thick cover of the main woods or the burrows which honeycomb their boundaries; and it is a good gun, well held, that stops eight out of the dozen in such circumstances. But we are getting our guns into the coppices—a culpable trespass, as they are reserved for a great occasion. Already a dozen or so of pheasants have met their fate, and the keeper's face begins to wear a dubious expression as he witnesses the untimely destruction of his *protegés*, when the sportsmen who have been peppering their ground game pull up a little reluctantly, emerging into the open once more.

If they get a chance of trying a willowy watercourse or some reed beds, so much the better, as game is found in such places now winter is close at hand that lends a variety to the bag. Walking quietly down the brook sides where the trees lean over, and the water rats make devious tracks in the soft mud amongst the arcades of pendent coral-red rootlets, they are sure to disturb something sooner or later. It may be a heron—that fowl that seems to have learnt preternatural shyness in the early centuries, when the sky was full of hawks and "pasties" full of his kind—a bird that rises half a mile away and sails up stream majestically; or, perhaps, it will be a fussy moorhen, as careless as the other is wary, that springs from under the foot and flutters away over the water meadows with both legs down—a comfortably easy victim, were we so minded.

But a duck is what we desire and hope for down here; and out yonder, where the stream has overrun a marshy corner and sown itself a garden of ragged watercress and kingcups, we may safely look to a find. The guns take either bank, and with the dogs in the slips we move quietly down. It is just such a feeding-place as the mallard loves. Amongst the deep soft growths the water has cut channels a foot or so deep, and runs quietly through them, rolling over the light gravel and playing about the jaws of the long, grey pike, lying silent in wait in the sub-aqueous glades for gudgeon and bleak which flash and frolic outside. A score or so partially submerged willows dot the swamp, and have collected amongst their branches tangles of reeds and grass on which coots build, and hassocks of water-rush—that whose pith the natives use for their "rush-lights"—mark the limits to which spring freshets rise. This is a sure find. The guns have just got within command of it when a mallard sails out from a floating cress-bed. He gives a look at the gunner on the right, floats round on the stream, glances at the other enemy, and is up into the air with a deal of splashing in a second. Two or three others rise and are accounted for very easily, and a bevy of teal may get up or steal away down stream, to be subsequently met with and thinned out as they circle round the gunners.

There is hardly time for lunch in these short days—the man in reasonably good condition ought to be able to walk and shoot all the brief hours the sun is above the horizon with little or no food. However, such exercise in the open air is an undoubted provoker of appetite, and we may suppose some sort of lunch is taken, leaning against a gate, perhaps, or seated on a fallen elm log, while the boys put out their game on the short, sharp-nibbled grass by the hedge side in a comely row, and report the total. Then at it again, beating the deep pits for rabbits, the broad hedgerows of blackthorn and briar for outlying pheasants, and picking up a couple or two of wood pigeons going to roost in the copper-trunked firs of the homestead, the sun sinking down meanwhile behind the western pastures, a huge globe of golden flame.

## The Quail Poacher and the Partridge Thief.

Just as anthropologists tell us that centuries of experience and toil intervened between the formation of the first smooth simple bone fish-hook, and the perfected instrument of capture armed with its fatal barb, so the learning of the field and stream has continually taught both savage and civilized races multitudinous and cunning methods whereby the beasts and fishes of forest and flood may be taken for food or other needs. Though this subject lays itself open perhaps to the charge of being one of "poaching," I am by no means ready to shrink before the spell of that dread word. Far too much "sport" with the gun is nothing but poaching at its worst, while the silent crafts of the woods so readily stigmatized, are often of higher science than the thoughtless suppose, needing a longer apprenticeship to qualify for success, and the possession of more self-control, more temper and judgment, with better eyesight and readiness than the "gunner" nowadays is called upon to display.

"The corn-land loving quail," as Drayton has it, may serve us as an instance of a game bird, eagerly trapped and snared wherever its presence is known and its culinary qualities appreciated. Those crates densely packed with unhappy victims that we see in the great poultry markets of the kingdom are usually from the warm shores of the Mediterranean. There the harvest of these small birds is one of the most important of the year, and men, women, and children have a busy time of it along the fertile shores of Sicily and up the changing Adriatic, capturing the quails in long nets as they arrive from the southward, boxing them in dark, shallow cases, where the faint light prevents them from indulging in the pugnacity of their species, and the low canvas roof puts a stop to the possibility of their courting suffocation by piling themselves three or four deep, or rubbing the feathers from their heads.

The plan of action is as follows. When the great annual migration sets in from the southward, and the flocks start on their long and dangerous journey from the African coast, the watchful "chasseurs" on the northern shores prepare a welcome for the wanderers which is more complete than kind. All along the edge of the tide, just above high-water mark, poles are erected at intervals of ten or twelve feet apart, and standing four feet from the ground. On the landward side of these slight notches are cut, in which rest the upper strands of long nets a little over three feet high, and often extending as far as half a mile along the brink of the sea.

Then commences the flight of the wandering birds, and some morning they begin to drop in by twos and threes, skimming swift and silent just over the surface of the water, all too "dead beat" with fatigue, doubtless, to keep a watch ahead, and so they dash headlong into the toils ashore, and the net falling from its notches keeps them securely prisoners in the meshes. These shore nets take males only for the first fortnight, as the females, detained possibly by family cares, do not migrate till later. The whole population are employed perpetually walking up and down the nets, replacing them where disturbed upon their notches, and packing the birds for the epicurean kitchens of Europe.

"It is a far cry to Lahore!"—or rather, to Cabul, in this instance; but the Afghans devote so much attention to our bird, and prize it so greatly, that their modes of capture—ingenuity itself—must not be overlooked.

Quails are caught, a recent writer informs us, principally with the object of securing the cock birds, which are used for combats. Quail and partridge fighting is as common in Turkestan as game fighting used to be in this country, and as goose fighting is at the present day in Russia. These fights attract crowds of natives, on which occasion a good deal of betting goes on, and a good, clever bird acquires celebrity. There are special bazaars held in the towns, which are much frequented, where young quails are sold, many of them bred from noted birds with a pedigree. There are two chief methods of catching quails resorted to by the natives. One is simplicity itself. A hair noose is fastened to a lump of clay, well worked together; a number of these appliances are scattered about the lucerne fields, which the quails are fond of frequenting; the bird caught in the noose is prevented from flying away owing to the weight of the clay, and its getting easily entangled in the grass.

The other method is more complicated. The sportsman has to represent an eagle; for this purpose he puts a stick tied in the form of a cross to another stick, which he holds in one hand, into the sleeves of his khalat—a long, loose overcoat; in the other hand he holds a net, similar to the one described above. Thus equipped, he proceeds to the field sown with lucerne or millet. There he begins to imitate the flight of the eagle. Walking along, he moves the stick with the khalat gently from side to side, or, stopping, begins to wave the khalat quicker and describing circles. The quail, seeing this terrible monster, tries to run away, but, being overtaken by the sham eagle, shrinks away somewhere into a hole or bush. The sportsman describes a few circles over the bird, and then secures him by covering him with the net. This kind of sport is also resorted to by the inhabitants of Samarcand.

Yet another method is with a kind of bag-shaped landing net. Early in the morning, the sportsman, accompanied by common house-dogs, which are trained to walk a few paces in front to start the game, enters the field, carrying the net, which has a long handle, holding it with both hands a little to the right. When the quail rises, the sportsman, by a very clever movement, covers the bird with the net, and then by giving the latter a rapid turn or twist, the game is secured. This, however, appears to be poor sport, six or seven birds being all that can be caught during an entire morning, and to achieve this even the sportsman must be well up to his work.

Again, Bellew says, in his "Journal of a Mission to Afghanistan," that in the early summer quails visit the corn fields and vineyards about Candahar in vast numbers; they are usually caught in a large net thrown over the standing corn at one end of the field, and are driven towards this by a noise produced by a rope being drawn over the corn from the other end, a man on each side of the field holding one end of it. When a quail has been beaten in fight his owner at once catches him up and screams in his ear. This is supposed to frighten the remembrance of his defeat out of his mind.

Beaumont and Fletcher, our earliest dramatists, allude to the "calling" on the familiar bird flute. With them the quail is represented as entering the nets of the fowler in response to the imitated cry of the female.

Several other striking modes of capturing these birds are practised in the East. One very simple plan is for the hunters to select a spot on which the quail are assembled and to ride or walk round them in a large circle, or rather in a constantly diminishing spiral. The birds are by this process driven closer and closer together, until at last they are packed in such masses that a net can be thrown over them and a great number captured in it.

Arab boys catch the quail in various traps and springes, the most ingenious of which is a kind of basket, the lid of which overbalances itself by the weight of the bird, much in the same way as that used in Russia for taking black-game.

To come westward again, and turning to the pages of that most popular of writers, the Rev. J. G. Wood, we read that in Northern Africa these birds are captured in a curious manner. As soon as notice is given that a flight of quails has settled, all the men of the village turn out with their great burnouses or cloaks. Making choice of some spot as a centre where a quantity of brushwood grows or is laid down, the men surround it on all sides, and move slowly towards it, spreading their cloaks on their outstretched hands, and flapping them like the wings of huge birds; indeed, when a man is seen from a distance performing this act he looks like a huge bat. As the men converge on the brushwood, the quails run to it for shelter, and creep under the treacherous shade. Still holding their cloaks on their outspread hands, the hunters, when within a few yards, rush upon the brushwood, flinging burnouses over it, and so enclosing the birds in a trap from which they cannot escape. Much care is necessary that the birds should not be driven in too quickly.

Bad as the Arab may be, negroes are amongst the worst enemies of game birds, both at home and all over the territory of the stars and stripes. In Maryland, for instance, they work havoc amongst the quail by means of falling log traps on the familiar figure-of-four principle, and in Texas a wail of wrath and anger goes up from the local gunners, who declare the negroes take entire flocks at a time, and they never set any of the captured birds free for stock. In this they show their characteristic improvidence, or want of regard for the future. The birds for the most part are taken alive to the neighbouring towns and villages, where they sell them at what they consider a big price. Their most destructive instrument is a mere pen built of sticks, and covered with brush. They have four trenches leading into the pen from opposite directions, coming to the surface about its centre. These trenches inside the pen are partly covered with bark and sticks, except at the centre, where they all come together. Corn or peas are scattered thickly in the pen and also in the trenches. When a flock of quail comes along, they find the food in the trenches, eagerly follow it up, and, with rare exceptions, every one of them goes into the pen, and is there a prisoner. He never thinks of looking down for the hole he came in at; he looks upward all the time and sees no way of escape. The freedman comes along and transfers the poor birds from the pen to his cage —from one prison to another. Thus whole regions are swept of their quail in the Southern States.

Darwin, in his delightful "Naturalist's Voyage Round the World in the *Beagle*," mentions a really sportsmanlike plan of action which, could it be imported, would add a pleasant variety to the list of English out-of-door sports. He describes how the South American Gauchos used to catch the tinamous with a small lasso, or running noose, made of the stem of an ostrich feather fastened to the end of a long stick. A boy on a quiet old horse, Darwin says, frequently would catch thirty or forty birds in a day.

This "tinamous" is a bird in appearance something between a quail and a partridge, but more closely allied in structure to the former.

It will be seen that the existence of these little birds, with enemies on every shore, is "not a happy one." Even in Malta the natives keep dogs who are especially trained for quail hunting. When the flights arrive the huntsman goes forth with his trusty dog and an ordinary casting-net over his arm. The dog hunts for the birds, which he finds by scent, and drives very slowly before him to the base of one of the numerous stone walls which cut up the island in every direction. Here the quail crouches, awed by its canine foe, who then remains motionless within a yard or so, until the man creeps up and throws the casting-net over both dog and bird.

In our own southern counties, advantage is taken of the disinclination of the quail to fly while it can use its legs, and V-shaped enclosures are formed with low brushwood sides, a couple of hundred yards wide at the mouth, but tapering down to a point, whence a single small opening gives access to a netted chamber. Men and boys then drive the game slowly into the open V, and the birds, running from their enemies, coast down the sides of the snare until they come to the apex, and crowd through the opening into the fatal chamber, the net over which renders their wings useless.

Turning to partridges we would suggest that, much as we owe agriculturally to the inventor of those highly perfected machines which shear our corn-fields to the last inch or two of straw, and lay out the yellow harvest with mathematical precision over the close-cropped land, from a sportsman's point of view the work is done too well and too thoroughly. Not so very long ago the occupation of the gleaner was a substantial reality, and less the utter myth of to-day, while the reaper's sickle or scythe never cut corn quite down to its last joint, but left a reasonable amount of growth, something that might afford fair harbourage for game. Now all this has changed, and leather leggings no longer brush through dry, ankle-deep stubble on the First, nor, saving under exceptional circumstances, or in remote districts where machinery is unknown, can keepers make the birds lie in the corn lands as they should without the aid of "hawks," or some other scarcely legitimate process. And not alone does the sportsman hear with qualified satisfaction the "reapers'" busy whirr, but there is also another class, the fraternity of nets and lurchers, who consider themselves aggrieved by modern economical farming, complaining that "birds" do not "busk" half so much on the stubble as they used to. Yet the poaching fraternity manage to find the coveys somewhere, and keep pace in ingenuity with the march of civilization, and, without doubt, most of those partridges which are to be had secretly a little before September dawns, or suspiciously early on the first day of that month, have come by their fate under the moonlight, untouched by any gun. One could wish that there was more room for the poacher to satisfy his love of woodcraft in these islands without infringing on the rights of private property, for he is at bottom generally a good fellow, and that instinctive love of the chase taking him to hedgerows with pockets full of cunningly twisted nooses, or a coil of close netting tucked away in the skirt of his coat, is in most cases essentially the same passion as prompts the enthusiasm and delight of the truest and strictest sportsman who ever pressed a trigger or shot over dogs. Nor must it be forgotten that the strict definition of poaching, like that of virtue itself, is apt to be modified and altered according to varying times and usages. What was lawful sport in one age may become the rankest sylvan high treason in the next; and Hodge and his kindred, who disregard strict limits of close seasons, or take any advantage of game which their wits can suggest, do no worse than many a keen fowler did a few generations back, greatly to the increasing of his own fame and advantage; so largely does custom regulate these matters. But the poacher of one kind or another has always been "in hot water." When those winged bulls of Nineveh were still uncarved in their native quarries game laws were, no doubt, an old and vexed question in the Assyrian courts, and mighty hunters of that empire exerted their skill in devising tortures wherewith to enhedge the sanctity of royal preserves and deter intruders. The game thief of our day need not fear having his eyelids cut off, or being buried up to his chin in the hot desert sands facing the sun, or sewn up in a raw bull's hide, which by gradual but resistless contraction crushes the life out of him. Yet these were once legal procedures for the same offence where Nimrod ruled. Even the milder but effective remonstrance of our Norman kings, the putting out of a right eye and the lopping of a right thumb in the case of those who misdirected their shafts amongst the king's deer, are measures too drastic for the spirit of the age. We do not now tie the hares Hodge has poached round his neck and make him wear them thus for a month or two, the evidence and punishment of his guilt, as was done with much success in mediæval princedoms, but we give Hodge "fourteen days" and bread and water, and even this does not cure him of his "delight on a shiny night."

Probably the least observant of travellers through our fair and fertile shires has noticed, as he has been swept by meadow and coppice on the iron road, withy bushes and brambles dotted about pasture and corn land, apparently aimlessly, where there could be no cause or reason for bushes to grow, and he will see they are not big enough for cattle to scratch against, and far too small for shelter. These, we regret to say, illustrate the watchful care required in modern game preserving, and the mistrust of all ungaitered kind abiding in the mind of the gamekeeper. They are put down wholly for the confusion of the poacher, and indicate the manner of that worthy's nightly raids. The partridge, unlike the pheasant, roosts upon the ground, choosing if possible a dry, elevated spot, such as a sandy meadow under a hanger, or bit of woodland sufficient to keep off the north wind. In such a spot the whole covey will collect at dusk, filling the vale with their pipings to gather the family and recall stragglers. Our poets have noticed this peaceful sound of the twilight. Burns speaks of "paitricks scraichin' loud at e'en," and Hurdis says, "I love to hear the cry of the night-loving partridge," while Grahame describes how at evening "stillness, heart-soothing, reigns, Save now and then the partridge's late call." But our poacher, as he sucks his short clay and leans on the weather-worn field gate, notes the soft sound, too, and not only the calling of one covey, but that of half a dozen, making his plans forthwith. Two or three men are required for this nefarious work, with a dogcart, if possible, to carry the spoil and facilitate escape. As soon as the pink glow is out of the sky in the east, and keepers may fairly be supposed to be enjoying an after-dinner smoke before setting forth for their evening patrol, the work begins, the trap is driven quietly to the scene of action, and while one hand stays by it to watch, two others take each the extreme ends of a long, fine drag-net, and walk slowly across the field, including, of course, all those spots where coveys have been marked down at sunset. That the men are loth to lose nets and gear when surprised, and will defend them with bludgeons or worse, may be understood when it is remembered these nets are sometimes of the finest silk thread throughout, as light and strong as they are costly. When birds are reached it is known by one or two rising tumultuously into the meshes; then the net is lowered instantly, and probably the whole covey, which sits close in a circle, tails inwards and heads out, is enclosed. Very short shrift then falls to the luckless brood, all of them finding their way before many minutes are out to the ready sack under the dog-cart seat, whereupon the net is ready again for another beat over the fallows or grass lands, provided they are not littered with those net-entangling bushes we have noted. "Poaching," Mr. Christopher Davies remarks, "is a terrible thing. It is even more fascinating than gambling, and in another way leads to as dire results. The desire of sport, the attractions of a life which is idle during the day, and busy only during the hours of darkness—the occasional large profits easily earned, the excitement of evading the law—make up a temptation which leads many a decent man to drink, misery, and crime; while his starving family has to be supported by the parish."

One species or another of partridge is found in every part of the inhabited globe, and everywhere they are eagerly sought after. We read, for instance, in Bellew's "Journal of a Mission to Afghanistan," how natives of Candahar adopt a very novel and successful method of enticing these birds within reach. They wear a mask or long veil of a coarse yellow cotton cloth, dotted all over with black spots, having eye-holes and hanging in loose folds round the body of the sportsman. Thus disguised, he creeps cautiously on hands and knees towards the spot from whence the "chickor" calls. The bird takes him for a leopard, an animal to which it has the greatest aversion, and will collect all its species in the neighbourhood with loud calls, and allow the make-belief to approach while they scream and flutter about him, when with gun or net, he can secure them with little difficulty.

For catching partridges, a peculiar kind of bow is used in Turkestan. It is formed of a long elastic rod, which is stuck into the ground, and then bent down and held in that position by a small catch arranged on a fork-shaped twig stuck into the ground; upon the catch are placed small sticks, on which the noose of the bow is ranged, under which some food is strewn, generally Indian corn. As soon as the bird steps on one of the sticks placed on the catch, the bow becomes detached, and he flies upwards with the latter, caught in the noose either by the leg or head. There is another original appliance for catching partridges in the same country, which may be called, for the want of a better term, a fowling line. It consists of a peg with strong twine attached to it, on the end of which is fastened a grain of Indian corn. A whole row of these pegs are driven into the ground along a path among the reeds. The partridge or pheasant in swallowing the grain, becomes captive, and remains attached to the peg, until the sportsman makes his appearance. These birds and smaller ones are taken by Pardis, a wandering tribe of Indians, in long, conical bag-nets, kept open by hoops, and provided with a pair of folding doors. Bullocks are used to walk through the long jungle grass, and drive the birds into the nets, without alarming them sufficiently to cause them to fly. After the usual thoughtless cruelty of their race, the Pardis break both wings and legs of each bird directly after capture, and thus the miserable victims are carried through a hot morning sun, all basketed together, to market.

Hardly less curious is a description we find in Johnson's "Indian Field Sports." The Hindoos are there said to equip themselves with a light framework of split bamboos, resembling the skeleton of a kite, and covered with green twigs, leaving two loopholes to see through, and another lower down for the insertion of the rod. This they fasten before them when they are in the act of catching birds, thus leaving both hands at liberty, and remaining completely concealed from view. The wand which they use is twenty-four feet long, resembling a fishing rod. They also carry with them horse-hair nooses of different sizes and strength, likewise birdlime, and a variety of calls, with which they can imitate the various birds' notes with the utmost nicety. As they proceed through the various covers, they use the different cries for the birds which they think reside there, and when the call is answered, suppose it be a bevy of quails, they continue piping them until they get quite close; they then arm the top of their rod with a feather smeared in birdlime, and pass it through the lower hole in their frame of ambush, and continue adding other parts until they have five or six out, which they use with great dexterity, finally touching one of the quails with the feather, which adheres to him. They then withdraw the rod, arm it again, and touch three or four more in the same manner, before they attempt to snare any.

In Nubia the village dogs, the lurchers of the White Nile poacher, are trained with much skill to run down and capture partridges alive in the furrows of the cultivated land. In Turkestan, sand-grouse, a species closely resembling English partridges in flight and habits, are taken by cotton drag-nets, worked almost exactly on the principle of those used by night along our own covert sides. These drag-nets are not the only methods of illegitimate sport our poachers know. In the eastern and southern counties of England, a peculiar sort of spring snare is used, set in the sandy spots of the fields, where partridges dust, whole coveys being taken thereby.

Sometimes, if the little bird is scared or wild, then a gang will take a change from stubble or plough, making a raid under the stars upon the woodlands. "On clear, bright winter nights," says a knowing authority, "when the full moon is almost at the zenith, and the definition of tree and bough in the flood of light seems to equal, if not to exceed, that of the noonday, some poaching used to be accomplished with the aid of a horsehair noose on the end of a long slender wand, the loop being insidiously slipped over the bird's head, usually a pheasant, while at roost. By constant practice a wonderful dexterity may be acquired at this trick. Men will snare almost any bird in broad moonlight." Pheasants are frequently taken by poachers in loops set in ditch bottoms and wood fences; but as the pheasant would probably withdraw his head were the noose made of wire, it is formed of plaited horsehair, and is then very successful. In fact, at home as abroad, "game" by no means sleeps secure until the shield of the law is formally withdrawn from it. Many a yellow stubble is swept, and many a coppice of hazel and larch visited before the lawful "season" begins, and all the care and vigilance of keepers is required to protect for "the First" the partridges or pheasants reared with that patience and costliness, which is a feature of modern game preserving.

# CHAPTER VII.

# *PIGEONS.*

### WOOD PIGEONS, AND THE CASE AGAINST THEM.

It must sometimes have occurred to any thoughtful person to wonder what heights the discontent of the British farmer would reach, were his mild northern plagues changed for some of those which scourge his fellow-subjects and kinsmen elsewhere!

He strains at a gnat in, let us say, the shape of a turnip-fly, while his brown-skinned brothers are swallowing that camel the locust; he grumbles profusely if an "emmet" or two gets into the dairy milk-pans, yet were he translated into a Hindoo grazier, dairy and cattle-sheds—I had almost added crockery itself—would crumble to dust before white ants; and what are cattle flies to tarantulas, scorpions, leeches, or mosquitos? Even pheasants swarming out to his barley fields are better than a score of wild pigs in a corn-croft; and, lastly, though our comparisons are not exhausted, what loss has he to complain of due to the amiable cushat compared to the havoc the passenger pigeon commits in America?

There he might be visited day after day by a solid column of birds "a mile broad, by two hundred and fifty miles long," as Wilson cheerfully expresses it; he might own their breeding-grounds, a wooded range of mountains where every spruce bent under the weight of nests, where the ground was white as though covered with snow for miles with the pigeons' droppings, and where the noise of wings when they rose at morning or settled in the evening was like the sound of a gigantic hailstorm on a frozen lake!

Compared to this our own native bird in its mild numbers is surely a friend. Once, perhaps, long ago, he was too numerous. Gilbert White writes, "I have consulted a sportsman who tells me that fifty or sixty years back, when the beechen woods were much more extensive than at present, the number of wood pigeons was astonishing; that he has often killed near twenty in a day; and that with a long fowling-piece he has shot seven or eight at a time on the wing as they came wheeling overhead. He moreover adds, which I was not aware of, that often there are amongst them little parties of small blue doves which he calls rockiers. The food of these numberless migrants was beechmast and some acorns."

To-day such gatherings as these are rare, as far as my knowledge goes. Small flights of ten or twenty, or more commonly still, a pair or two at a time, are about the usual numbers visiting our enclosures and fields. Wherever woods are, there will always be some, though they fly long distances for food.

Our English poets, who are deeply indebted to the pigeon for a thousand metaphors, seem to have been quaintly "at sea" with regard to the varieties of their favourite bird. "Apart from the dove general," Mr. Phil. Robinson tells us "the poets employ the dove particular—the ring dove, the stock dove, and the turtle dove. But what relation each species bears to the other the poets never considered themselves at liberty to determine. Watts makes 'the turtle' the opposite sex of 'the dove'—'no more the turtle leaves the dove'—but allows at the same time by implication the existence of a female turtle; while Cowper makes it the female, though elsewhere, with Spenser, making it the male. Thomson uses the stock dove as the male of the turtle, Cowper as the male of the ring dove, and Wordsworth as the female of it. As a general rule, ring doves are 'he' and turtles 'she' (chiefly widows); while stock doves are one or the other, as poetical exigencies require. But the ultimate outcome of this reciprocity of sexes and species is a ring—stock—turtle—dove, as elastic in its properties as even poets could desire, and as variously endowed as any Pandora-Proteus."

To return to stern facts again, farmers shoot the ring dove when they can; firstly, because he is fond of peas, and secondly, because he loves turnip tops, especially when the weather is hard.

It would take more amiable effrontery than I possess to deny these charges. Pulse of every sort or kind has an irresistible attraction to Columbian nature, and is searched for eagerly and devoured greedily wherever it is obtainable. While there can thus be no doubt that wood pigeons will eat all the peas or tares they can find, I am not quite sure, from my own observations, whether they actually shell them for themselves. If they cannot and do not, then half their guilt is purged at once, for it is obvious that the consumption of shed peas—no good to any one—is a very light offence.

In the turnip fields they graze like a flock of sheep, tugging the tender green leaves from swedes and "Carter's best Dutch" roots preparatory to bolting them in pieces as big as a postage stamp. I have shot them when they have been returning to roost from these vegetarian excesses, and on several occasions a bird's crop has been so full of this food that it has burst in falling, and a large handful of leaves has been scattered about. Surely the gastric juices which can assimilate such a mass of raw stuff during the hours of the night ought to be the envy of all dyspeptics. The birds, however, seem to thrive on this diet, while their flesh takes a rather strong and musty smell which can readily be recognized after a little acquaintance.

Were these two items the only ones on their bill of fare, the case against the wood pigeons would indeed look serious. But it is not so, and any one interested in the question should shoot a bird or two when they seem to have been doing the greatest mischief, and make an examination into the uncontrovertible evidence of their crops. The revelation will, I venture to think, be instructive. There is a passage in Mr. St. John's "Sketches of the Highlands" well illustrating this. He writes—

"An agricultural friend of mine pointed out to me the other day an immense flock of wood pigeons busily at work in a field of clover which had been under barley the last season. 'There,' he said, 'you constantly tell me every bird does more good than harm; what good are those birds doing to my young clover?' On this, in furtherance of my favourite axiom that every wild animal is of some service to man, I determined to shoot some of the birds to see what they were actually feeding upon, for I did not at all fall in with my friend's idea that they were gorging upon his clover.

"By watching in their line of flight from the field to the woods, and sending a man round to drive them off the clover, I managed to kill eight of the birds as they flew over my head. I took them to his house and we opened their crops to see what they contained. Every pigeon's crop was as full as it could possibly be of two of the worst weeds in the country, the wild mustard ('charlock') and the ragweed, which they had found remaining on the ground, these plants ripening and dropping their seeds before the corn is cut.

"Then no amount of human labour and research could collect on the same ground, at that time of year, even as much of these seeds as was consumed by each of these five or six hundred wood pigeons daily for five or six weeks together."

The above well indicates the importance of condemning no bird on appearances. Without such practical evidence the farmer would pay a boy daily to scare away the doves, would pay again to hoe out the charlock and ragweed before burning it, and lose once more at harvest by a "dirty" sample of corn, *and every penny of this would have been wilfully thrown away!*

Wild fruits figure largely in cushats' meals. From a crop recently opened, six hundred and ten ivy berries and a few undistinguishable fragments were taken. The Rev. F. O. Morris, in his "British Birds," points out this diversity. "The wood pigeon feeds on grain in all its stages, wheat, barley, and oats; peas, beans, vetches, and acorns; beech mast, the seeds of fir cones, wild mustard, charlock, ragweed, and other seeds; green clover, grasses, small esculent roots, ivy and other berries, and in winter on turnip leaves—and their roots in hard weather—the first-named are swallowed whole.

"It may safely be said that any damage it does, and it must be confessed some is done by it amongst seed tares and pea fields, is abundantly compensated by the good it effects in the destruction of the seeds of injurious plants."

We do not think there is anything more to be noted in this subject. It is plain that in excessive numbers and in certain districts the "quist" might become very harmful to one or two specialities of agriculture; but what has been said should indicate the bird has a usefulness of its own. One other charge against him we are bound to notice. It is one that chiefly affects the male bird and not that luckless waive, his mate.

Wood pigeons, as well as blackgame, and the rapidly increasing capercailzie in certain parts of the country, do damage to woods. The two latter feed upon the young shoots of firs and other trees, and buds and twigs may be turned out of their crops sometimes. Wood pigeons do harm in a different way. Any one who has walked through woods frequented by them, about five o'clock on a bright summer morning, has doubtless been soothed by the cooing of unnumbered doves perched on the tree-tops, their mates keeping house below. A beautiful sound it is, but that is just the time the mischief is done. Every pigeon sits as near heaven on a fine morning as he can, which, in a fir wood, means on the leaders of the trees—at that season young and tender. In the leader of a young fir is centred all the promise of a clean, straight stem, pointing direct from the axis of the earth to the zenith. When this is bent aside, the tendency, especially of all the *Picea* or silver fir tribe, is to send up several leaders; the result being seen in double or treble stems, instead of one fair clean shaft. Therefore the wood pigeons, which are less easily kept under control than rabbits, may be looked on as injurious to young fir woods to some small extent. But it is not every fir they perch upon even in a forest of larch and spruce, nor can we think any but a very small proportion of these resinous little apexes would be permanently distorted by their weight. More often plantations are of several sorts of trees, the beech and aspen overtopping the firs, at least while the latter are young and tender. The highest trees, again, in a wood are often the stunted, worthless little bushes that crown some rocky knoll clothed with fern and foxglove. If a legion of pigeons were to perch on such a spot for a month, the damage done would not amount to the value of the good white paper wasted by him who first made the accusation! I cannot think this indictment is a very important one.

And then what a pleasant bird the wood pigeon is, and surely of more account than many timber merchants! Coppice and hanger would lose half their attraction without his presence, the loud beat of his wings as he takes to flight, or the flash of his blue plumage where the sun comes down through the branches. Truly the ring dove is not much of a game bird, though I have stalked him when I first began to shoot with all the patient ardour of a Red Indian, and held him a well-earned trophy at the end of an hour's watch.

He is most off his guard—and I betray him with some reluctance—when returning at night to his fir trees, and by standing quietly beneath them as the birds circle round and drop down into the deep shadows, many a gallant pie may be furnished. But such sport should not be disguised under a thin veneer of virtue. We shoot the wood doves because they are toothsome; they are of no more harm in the aggregate to the farmer, I confidently believe, than "the tame villatic fowl"—and not so much, if the enemies of poultry are to be believed.

In winter time there is hardly a bird in the country side which makes its presence more felt than this watchful and suspicious admirer of pea stubbles. Pheasants have been decimated, the partridges scattered, and coverts beaten as much as they will be each year, yet the wood pigeons are as numerous as ever. Gamekeepers have an inveterate grudge against them, and farmers, at least of the old unreasoning school, attribute numberless enormities to them. For them to suppose they will fly over a ripe pea-field and not perform a couple of turns round and then descend to take toll of the tempting pulse, is to expect too much, as I have said. But peas are ripe before pigeons congregate, and a pint or two pillaged by them counts but little in comparison with the pecks of the seed of gaudy but villainous charlock, and of flaunting poppy grains they make away with. That they enjoy turnip-tops in hard weather there is no denying; but those holes pecked into the roots themselves, which let in frost and do much damage, are not done by the quists. Therefore, against the agriculturist's condemnation of these dwellers amongst the beech trees, we put forward the familiar Hibernian plea that "the culprits are innocent, and moreover have extenuating circumstances in their favour!"

An examination of a few birds' crops would always restore them to rustic favour, and Master Tommy or Harry, fresh from school, will undertake the obtaining of the birds with delight. Indeed, waiting as they come to roost in their favourite ivy-covered firs as mentioned, is an amusement not without its pleasures for those who shun the "pomp and circumstance" of modern sport. There are the long vistas of the pine stems glowing red in the last rays of the winter sun, the yellow and scarlet carpet of last autumn's dead ferns below, and the bits of blue sky overhead across which flash the birds destined to bear posthumous testimony to this argument of ours. Or we may wait for them in the day time amongst the decoys, with success proportionate to our knowledge of their habits and powers of hiding.

Of four species of doves inhabiting Britain there is only one really known to a non-ornithological public, or familiar to those growers of corn and roots who naturally regard all birds from the standpoint of their usefulness or destructiveness. This bird is the blue-rock, the cushat of the poets, the quist, wood pigeon, and ring dove of country people. About those turtle doves that come with the spring from Algerian and Spanish olive gardens, the grower of swedes and barley need not trouble himself. Country ramblers in Kent or the Midlands, however, who keep their eyes open, may have noticed wisps of these birds, in appearance a little like missel-thrushes, with an extra allowance of tail, but characterized by a true Columbian flight, rising from pea stubbles or open stony glebes, where their colour matches so exactly the surrounding wastes that they are invisible to man or hawk until they rise on the wing. But these "Wrekin doves," as they call them in Yorkshire, are not numerous anywhere, and still less conspicuous even in the select localities to which they return year after year. Probably few other birds of their size in England are less molested. Hardly ever shot at except by a young and radical gamekeeper, or an early partridge shooter of an inquisitive turn of mind, they use our woodlands for their nestings, and get away again southward while hedgerows are thick and the yellow autumn corn still nods to the south-westers.

The stock dove, another of our four species, is very generally confounded with the common blue pigeon by those who ought to know better, though it must be conceded there is much resemblance between them at a little distance. Near at hand we may recognize the former by its lesser size, a different shade of colour, and the absence of that white ring round the neck which marks a true wood pigeon. This bird is perhaps less a dove of the high woods than the cushat; it loves outskirts and open warrens, where, as often as not, it utilizes a deserted rabbit burrow for a nesting-place—a curious fancy for a pigeon!—and deposits two white eggs an arm's length down amongst black roots of bracken and wire-tough fibres of ling. Both these latter species, feeding together, are often included in the same sweep of a fowler's net and tumbled incontinently into his market crate, while many worthy folk who see them on the poulterer's hooks suspect no difference between their breed and that of the ordinary pigeon of commerce.

The rock dove of St. Abb's Head and the caverns of the Cornish coast is the last English bird of this family. But we have now in view some means, besides that of the gun, by which wood pigeons may be induced to leave our tender young turnip-tops alone, or may be checked in their larcenous enterprises with relation to the expensive food we put out in our woodland drives for the pheasants during the hard weather. Amongst green crops, trapping the pigeons is probably the best remedy to be adopted. If a few common gins be set in the neighbourhood of the spots at which the pigeons mostly congregate, birds are certain to be caught, and so alarm the others that they will not return for some time. Stretching pieces of stout cotton, to which feathers or bits of red flannel are attached, is also very startling to these fowl. For the open spaces in coverts, or in a drive or clearing under trees where pigeons perch, a long, strong, and springy ash or other pole, of about the thickness of a man's wrist, is sometimes securely fastened down by one end; the other loose end is then drawn as far back as possible, and held there by a peg so placed as to be easily withdrawn by a string held in a distant hiding-place. After a few days' feeding with pheasant food, the wood pigeons will come in great numbers. Food is scattered on the space which will be swept by the ash stick when released, and the watcher then retires to his secure hiding-place with the end of the string in hand. On a sufficient number of birds having settled, the string is pulled, the peg is withdrawn, and the sapling flies round, carrying destruction to everything in its path. The great advantage of this villainous device is its silence.

Were the English "velveteens" less conservative and orthodox in his views of what the limits of his duties are, he might take a hint from the "foreigner" in trapping blue rocks. The woods in Northern Italy are often bisected by narrow and deep road-cuttings made by the charcoal burners and others. Pigeons are taken here in vast numbers by a method which must be seen to be fully understood. From tree to tree in the road-cutting a light but strong net will be hung. Small boys then take up their position on stages built among the branches of neighbouring trees, and whistle and call as the birds are returning to roost. When a flock approaches, such a boy whirls round his head a stuffed pigeon, having a weight in its head and a string near the tail, by which he holds it and hurls it at the net. The wild birds, accepting its treacherous guidance, swoop down and dash into the net, in which they are at once entangled. A hundred or more will be taken at once by this device. It is, perhaps, too much to expect of the guardians of our woods and coppices that they should perch themselves in the fork of a convenient oak and there twirl the disc and drop the net as the raiders of corn shocks and pea haulm go homewards down chestnut-covered pathways after their foraging expeditions. Yet, where pigeons migrate at certain seasons in large bodies, good work is done by a method nearly allied to that above described.

If we suppose ourselves standing in a gap on the sky-line of the Pyrenees or Savoy Alps at daybreak, we shall see how the mountain herdsmen replenish the village markets and provide the chief ingredient of pigeon *pâté* for wandering tourists. On the very ridge there is probably a little level ground—an inviting pass between rocky bush-covered crags on either side. This little plateau is clear of underwood, and three or four oak trees, left at convenient distances, dot its surface.

It serves as a tempting and well-used short cut for the pigeons coming down the valley; and among the branches of the oaks through which they would naturally pass are planted long, strong poles, supporting nets reaching to the ground, and so arranged with "bridles" and pulleys that they can be made to collapse instantly, one after the other, from neighbouring hiding-places. Should any one think of trying this arrangement in Canada or elsewhere, he may be able to do so from the more detailed description of a writer in the *Field*. He describes the ridge as "more or less level for some two hundred and fifty yards, which space has been cleared of all wood with the exception of six huge oaks standing in line, but rises abruptly from the level to the eastward. At the height of about forty feet in each oak was fixed a spar, from which depended a rope, with the lower end pegged to the ground, and carrying a wooden travelling ring weighted with iron. Each spar also had a block and halyards, the standing part of the latter being fast to the wooden ring. The nets, one inch and three quarters mesh, and about fifty feet broad, have their upper corners hooked on to two of the wooden rings, and are thus hoisted into position; the lower ends are drawn backwards, *i.e.* southwards, for about thirty feet and pegged down; the two halyards of each net are hooked to a single trigger, and all is then ready."

On commanding points overlooking the glen that lies in the purple shadow of daybreak are posted small boys, whose duty it is to keep the pigeons to the valley, and warn the netsmen of their approach by a little judicious shouting. On the ridge the leader of the gang perches himself in a tree a little in front of the nets. He is armed with some light wooden discs, which he hurls at the birds as they pass him, with the result that they swoop down and come within the drop of the nets. Waiting in "the chill of the pearly dawn" is no doubt cold enough in these mountain solitudes; but as the sun comes over the ranges in the east, a traveller tells us, "we begin to hear the cries of the flagboys in the distance announcing that the first of the birds were in sight, and the fears entertained of a possibly blank day are dispelled. In a couple of minutes a shrill whistle from the chief was the signal for every one to rush into hiding. A few seconds of breathless suspense, and the silence was broken by a rushing sound overhead and a simultaneous collapse of four nets with seventeen blue rocks fluttering on the ground underneath them. It was not, perhaps, sport, but it certainly was most exciting. The net men rushed out and retrieved them. Each bird as it was gathered was plucked of the feathers of one wing, and put into the front pocket of a sort of apron the net men wore, and eventually transferred to a receptacle formed of boughs built round the trunk of a tree. As soon as the last bird was gathered, the nets were smartly hoisted again, as the shouts of the flagboys were already heard. In a few minutes, another whistle and another rushing of wings; but, instead of the rattle of falling nets, there ensues a perfect hurricane of the most awful oaths from the nest in the beech tree, proclaiming to the initiated that the pigeons had passed over the nets and gone on their way untouched." Twenty to thirty dozen birds are sometimes taken in the early twilight by this curious and unique arrangement.

The beautiful fruit pigeon of Bengal, brilliant in yellow and claret colour, is taken in nets hung between fig trees to ensnare flying foxes. These nets, or something very like them, are mentioned in the oldest Sanskrit writings, and are suggested on Assyrian bas-reliefs and Egyptian frescoes. The American passenger pigeon, a genuine farm pest, is thus caught in Maine, U.S. A piece of ground, about thirty feet by fifteen feet, is levelled and smoothed so as to look something like a concrete tennis court. Near it are fixed some tree tops, about twelve feet to fifteen feet high, forming a sort of fence around, and affording a place for the birds to perch on. A bough house or cachet is built at one end of the bed, in which the catcher could sit hidden from the doves. Indian corn is put on the earthen floor, as it is called, every day, until the birds come to find it out and come to feed there regularly. The American pigeons turn up in flocks about four p.m., and in the morning. When it is known, by watching from a distance, that they feed, the fowler goes down about a couple of hours before and sets his nets, and sits in the bough-house until they are together, and catches the lot by pulling the net at a judicious moment. Very likely this plan, with a little modification, might be very successful amongst the faggot stacks and clearings of our beech and oak woods.

To the poisoning of birds or animals of any sort there is always the greatest objection. Not only must all such methods seem criminal and cowardly, but they are often absolutely dangerous to man and beast. Only a short time ago the papers told us how a farm bailiff in East Kent had, together with his wife and family, had a narrow escape from death by arsenic. The man was walking through a wood with his master, when he picked up a wood pigeon, apparently freshly shot. He took it home and had it cooked. A few hours after they had partaken of the meal he and his family were seized with illness; and the man himself showed such serious symptoms that a doctor was summoned. The usual remedies in cases of poisoning were administered, and the man recovered. Another device, perhaps one degree better than actual poisoning, is to place a sheaf or two of oats or wheat in the field, and allow the birds to freely feed from it for a day or so. Meanwhile some grain must be well soaked in gin or brandy, the commoner and more fiery the better. This should be spread thickly round the sheaf or sheaves, and the birds coming to feed upon it are quickly stupefied and easily caught. A day's watching of such an arrangement should result in a pretty fair clearance of birds. Our own wild pigeons are not to be decoyed into the snare by means of their own species, as is the case with some kinds. Yet curiosity, or the suggestion of safety and food which the presence of their kind at any spot seems to denote, will often bring them within range of the gun, which, after all, is the most apt instrument of destruction we have for them. The worst of it is, that they are so superlatively keen that the slightest error of judgment in making or placing the "stales" is fatal to all chance of their usefulness. Syrians call down flocks of wild doves by means of a tame decoy, whose eyelids are sewn together, and who is fastened in the neighbourhood of the hidden gunner armed with his long ancestral matchlock. In the same way, in China, a dove is fastened, blind, at the end of a slender switch jutting out from the top of a tall pole planted in the ground. The bird's weight causes the wand to bend and swing continually, thus obliging it to use its wings in order to preserve its balance. The decoy's movements very effectually attract the notice of roaming flocks, which soon alight all around the screen behind which the Celestial lies in wait.

Outside our coppices the village gunner uses decoys of wood, metal, and indiarubber, shaped and painted to resemble the live birds as nearly as possible; but, as has been said, there is an art in the placing of them, and even when that is mastered we doubt whether "the game is worth the candle." A gunner who knows the habits of the bird will probably manage to pick up quite as many pigeons without allies as he would with the best of them.

# CHAPTER VIII.

## *DUCKS.*

DUCKS IN MARSH AND MARKET—THEIR ECONOMY AND FUTURE.

No doubt the patriotic wildfowler is glad to see agriculture creeping down to every sea shore, and a careful husbandry snatching from tidal rivers and estuaries, foreshores and marshes, since the steady advance of coulter and mattock indicate the national vigour. Yet he may be excused a sigh, for there cannot be the smallest doubt that wild-fowling is becoming harder and harder to obtain every year, at least in the southern half of England.

The state of the case is familiar enough to every marsh-man. It matters little where he turns his steps, the old familiar happy hunting grounds exist no more, as far as sport is concerned; the wastes inside the land wall have been ploughed up, sown with lime, and now a trim crop of turnips or button onions for some manufacturer of pickles grow where, a little time ago, when our first gun was still in its brilliant newness, the dykes were open and knee-deep in water, blackthorn and elder formed impenetrable thickets, ruffs played on the hummocks, big, wild-looking cattle enjoyed their wallowings amongst whispering rushes shoulder high, and nothing was heard but the larks and plaintive whistling of plovers. But of course ducks don't care for the turnip furrows or carrot plots, and they have gone with the rest of the wild fauna. They will go just as certainly if a series of tall volcanic chimneys soar into the sky, while enterprising chalk works or what not "realize the natural wealth of the neighbourhood;" and thus the wildfowler's land is all in new hands, the rivers on which he punted to widgeon or brent geese are turgid and churned by the screws of unnumbered steamers. Along the actual sea coast it is nearly as bad. There used to be a strip of debatable land, saltings that the sea, the shore-shooter, and herdsmen shared equally between them; but now the latter seems to have monopolized them. He keeps the sea out with dreary mud banks, and the shooter with notice boards and well "spiked" gates!

This is the unacred sportsman's view of the matter. For those who own some soil by the water side, or even a pond or two where wildfowl fly in of a night from the open sea, the case is not so bad. They may still know a scaup from a scoter, a shieldrake from a shoveller, when they come across them. They shoot for their own larder, and when the flight is on, or the weather severe, there are still enough stray birds about what were once our wild lands to satisfy all their modest demands. But profitable "decoy ponds" are things of the past, and some of our largest game salesmen, with whom I have gossiped on the subject, say we are more and more dependent on Holland and the German coasts for our wildfowl supplies. For a single week of January, 1886, we drew "fur and feather" to the value of £15,000 from across the North Sea, and meagre indeed would be our market stalls were this source of supply to fail! Of course our seas will always attract wildfowl, the great wilderness of the Scottish kingdom, and the vast bays and sheltered estuaries of the Green Island especially must remain more or less productive. Sir Ralph Payne-Gallwey puts down the yearly bag of the Wexford puntsmen, some dozen or so in number, at three or four hundred birds apiece, even in the recent succession of mild winters that have characterized our climate, and he has seen three or four thousand widgeon on a single sheet of water! I do not think there is any prospect of these *outer* wildfowl grounds being depopulated to any visible extent for a long time, the breeding grounds in the far north are so immense, and the position of England is so admirable for attracting migrants. No doubt the fowl will become educated to a high pitch of suspicion as they are more and more sought after, our gunmakers being called upon to meet the emergency by still more powerful fowling ordnance; but I *do* think our inland wildfowl resources are neglected and jeopardized; with a little attention they might produce far more profitably. It is too much to expect, perhaps, that more decoy ponds shall be started for tempting teal and ducks to breed with us; but I think we might give some of our shires a better repute with the wandering feathered tribes by a little skilful management.

The Wild Birds' Protection Act has certainly been a step in the right direction, and has been of considerable benefit to our birds already. The close of the shooting season, April 1st, is rather late, doubtless, for many species of ducks. Partridges then take wing from the hedgerows and spinneys silently in couples, and without all that bluster that has marked their flight since the last broods were disposed of; while wild ducks are put up in pairs from the river edges, and have for a fortnight or so, we suspect, been deliberating on the momentous nesting question, "Where shall we go?" They ought to have been allowed to deliberate in perfect peace.

Good, too, has been done by the admirable enlightenment of many noble and extensive landowners, who have listened to the teachings of Waterton and given our wonderfully rich bird fauna a home and sanction in their coverts and lakes.

Others, again, who are not amongst the "acre-ocracy," love and study the many forms of life along country sides or sea shores with the amiable generosity and intelligence of a Gilbert White. The century and its liberal teaching has even produced a naturalist-gamekeeper or two, and I have read with wonder and delight in country papers keen and intelligent observations from the pens of those who twenty years ago regarded their mission as simply one of slaughter. On all such rests the best hope of our rarer species. If the magnates of the soil, like Lord Clifton, of Cobham, would issue a general Order of Amnesty, if reeves and bailiffs would observe, and if naturalists would use their field-glasses in preference to guns, we might add fifty kinds of birds to our common species, all of which should delight many and harm no one.

Flapper shooting (though I have shot flappers enough myself) is a very doubtful legitimate sport, and it is impossible not to suspect it gives a neighbourhood a bad odour in the minds of the survivors who return, at least for a time, as salmon do, to their first nurseries, and would, if all were promising, and there were no unpleasant memories, undoubtedly use them again. There are plenty of odd corners in water meadows and by stream sides where ducks would stay and breed, if the crow boy with his gun were suppressed, and they found peace and a little shelter. Such shelter they might well get from osier beds dotted down, or low waste lands. These osiers are in themselves a profitable crop—we imported five thousand tons of them last year from our sagacious neighbours across the Channel—and they fetch £8 per ton, nearly three times as much as good potatoes, and eight times as much as the roots the farmer cultivates so carefully for his cattle. Moreover, though they grow best by water, much water is not always essential to them.

It may tempt some attention to this matter if I add a recent letter of a correspondent in the *Field*. He writes: "The osier has been cultivated here in Norfolk with great convenience and profit for some dozen years. I once got some shrubs from a well-known nursery, and when unpacking these was struck with the extraordinary toughness of the "withy" bands with which the bundle was girded. A set was cut from the least bruised part of the band, and stuck in between two of the plants it had enclosed. It took root, grew, and in the next spring, from the four shoots it had made, four more sets were inserted among the shrubs. These, too, grew vigorously, and finding, what was not expected, that osiers will thrive far apart from water (the well is ninety feet deep, and running water is not within a mile), sets having been thrust in everywhere among the shrubs, among the underwood in little plantations, between small trees in orchards, in spare corners all over the place; and, except where gravel comes near the surface, the osier grows vigorously. But the soil is a good loam on a brick earth. The advantage of having an ample supply of 'bonds' to tie up faggots of all kinds, to bind faggots (when tied) to rails, so as to make stockyards warm, or fit up temporary places of shelter, and to fasten up bundles and hampers, is very great; many balls of cord are saved. Every February a man goes round and cuts all the osier stools down to the stump, and where the neighbouring plants are ready to occupy all the ground, roots out the osiers which have served as nurses and temporarily occupied part of the soil. Besides those used on the place, there are generally four or five bundles which are sent to the neighbouring basket-maker, who weaves them into hampers, fowl baskets, and into what articles of wicker-work are wanted, making two sizes, so as to use all-sized twigs. The variety has a yellowish-brown bark, a smooth shining leaf, and is quite free from any kind of efflorescence. It does not seem to differ when worked up from the ordinary appearance of unpeeled baskets, so probably the osier is the common variety; yet it has won itself a local reputation here, and some twenty or more people have come to ask 'cuttings of them tough bonds of yourn.' This osier grows, but does not thrive, on sand, gravel, or a bank; on the flat it grows vigorously."

Here is a chance whereby enterprise may pay the rent, utilize waste marshy corners, and afford cover and hiding-places for several sorts of wildfowl!

It has been ingeniously suggested that the direction of the prevailing wind in their remote summer quarters, when our winter wildfowl rise to go south, may somewhat influence their abundance or scarcity in certain localities, during the next four months. This is plausible enough, and might explain why we have periods of abundance, and others of scarcity, without much seeming regard to the ruling of the weather. The whole subject of migration, and the laws which govern it, are as yet imperfectly understood. They form a rich field for working naturalists, who would find much information collected ready for their use.

In conclusion, I think that while our admirable insular position will always assure us a fair portion *on our seas* of whatever game is afoot (or rather on the wing) in Europe, our inland waterfowl resources yet require to be husbanded, if the mallard and the teal, with their curious and various kindred, are not to be banished to remote Irish bogs and inaccessible highland tarns.

There are other dangers for the ducks on these shores besides those with which the over-covetous gunner threatens them. The decoying of the whole tribe is a curious art in itself.

## Duck Decoys and Devices.

It is just at the winter season of the year that the wildfowler's hopes are at the highest—whether he be the amateur, floating over the saltings in his new punt, anxious to try the range and scatter of a big gun from Holland's; the professional with weather-worn, but none the less deadly, gear, or the fen man of nets and decoys—each and all watch the weather intently while meditating on the prospects of a good winter's bag of wild duck, widgeon, teal, whichever their locality best produces. The puntsman's and shore-shooter's pastimes are well understood, but there is more excitement and variety about the decoy man's fashion of bringing his game to book than many people know. This is chiefly because decoys and flight ponds are rare, having ceased to be profitable establishments to the man who works them for mere money gain in a majority of instances. And sportsmen of the old school who could take an honest interest in and enjoy the management and working of a well-frequented pond are, it seems, becoming scarcer and scarcer. There is not one sporting estate in a hundred where a decoy can be seen at the present time, though good and convenient sheets of water are numerous. Two chief kinds of decoys are used: the first, for pochard, is called a flight pond, and has nets fastened to tall, stout poles, twenty-eight or thirty feet long, round its margin. At the bottom of each pole is fixed a box filled with sufficiently heavy stones to elevate the poles and nets the instant an iron peg is withdrawn, which retains the nets and poles flat upon the reeds, small willow boughs, or furze. Within the nets are small pens made of reeds three or four feet high, for the reception of the birds that strike against the nets and fall down. Such is the form and shortness of the wings of the pochard, that they cannot ascend again from these little enclosures. When all is ready, the dun-birds are roused from their pond, and as all wild fowl rise against the wind, the poles in that quarter are unpinned, flying up with the nets at the instant the birds begin to leave the water; they are thus beaten down by scores. This is not, perhaps, a proceeding which gives much opportunity for the display of great skill or science, certainly ranking below the decoy proper, wherein wild fowl are enticed up a covered "fleet," the utmost caution and care being required from first to last to prevent them from becoming suspicious or doubling back on their captors.

Mr. Christopher Davies, who has just published a delightful little volume for boys, entitled "Peter Penniless, Gamekeeper and Gentleman," thus happily describes the appearance of the pond. He says: "They were now in a great bay, which was as secluded as it is possible to imagine. The thick wood overhung the water, then came a bed of reeds, then a stretch of water-lilies. An arch of bent saplings spanning a dyke was but the commencement of a sort of network tunnel, about ten feet high, and eighteen broad at the mouth, but gradually narrowing and decreasing in height, until, at the end, it was only about two feet in diameter. The last ten feet of it were detachable, being formed of network stretched on hoops. The dyke over which the pipe was erected, was very shallow, and, of course, narrowed as the network did. It was about ninety yards long, and was not straight, but curved from the lake to the right for a quarter of a circle, so that when you were at one end of it, the other was not visible. There were high banks on each side, partly natural and partly artificial, and thickly clothed with underwood, and the outer side of the curve, which was the one from which the decoy was worked, was screened off from the pipe by a series of reed screens or fences, placed diagonally with their broad sides inclined towards the lake and overlapping each other. Thus, any person approaching the pipe in the proper manner would be perfectly invisible from the lake, but would be able to see up the pipe with ease. The screens were connected with each other by lower cross fences called 'dog jumps.'"

Ducks of all kinds feed chiefly at night, and fly abroad for that purpose to pools, marshes, estuaries, and other likely feeding-places, returning at daybreak to the quietest and most sequestered lake they can find, where they sleep, rest, and preen themselves during the day. Now, a decoy-pond is designed to give them the absolute secrecy, quiet, and rest which they like. Here they are never disturbed, even by the destruction of hundreds of their companions, for the decoying is carried on with so much secrecy and quiet that, if a score of ducks are having their necks wrung at the funnel of the pipe, the flock of fowl on the water not a hundred yards away are blissfully ignorant of anything unusual happening. As night falls, the ducks fly away to their feeding, and this is called "the rising of the decoy." At dawn they come back again. The pipes or lake must not be approached in the day time, save for the purpose of working them, and all the work which has to be done in clearing out the dyke, repairing the net, laying down food—barley or corn—in the shallow bay, breaking the ice, and so on, must be done at night. In times past, two to three thousand birds was a good bag for the season from one pond's working, now fifteen hundred would probably be all that could be looked for from the same lake, if so much. Now, let us see how the complicated machine works. We go forth on a clear, fresh winter afternoon, such as—in spite of the abuse heaped upon it—the English climate affords us now and again, with a keeper and an eccentric dog, of foxey yellow hue, silent and obedient in habit—an unobtrusive but all-important member of the party. For some distance the path is, perhaps, over furze-covered downs within sight of the sea, which lies, a dull leaden sheet, a mile or two away under the low red winter sun. Then the track enters the woodlands, and leads by the side of a trickling stream, under hazel bushes, already tasselled with green catkins in preparation for spring, which comes nowhere earlier than to these sheltered hollows, and so up by mossy slopes and yellow-ferned dells, to where a ring of willow trees show their characteristic outline against the sky. Here the keeper insists upon absolute silence, perhaps handing the spectator of what is to follow, a smouldering brick of peat upon which he is instructed to breathe, and so obliterate his personality to the keen-scented wild fowl, the man taking another himself. Then commences a cautious approach to where a five-foot fence of reed or wattle shuts out the lake that lies beyond. This reached in the most perfect silence, not a twig having been broken under foot, they make themselves a spy-hole and peep through. The water, some four or five acres in extent, is dotted all over with fowl feeding and cleaning themselves, and close by are "a company" of widgeon, "a lord" of mallards, "a badelynge" of ducks, as old-fashioned fowlers had it, or probably some of each kind, in a gay and busy crowd fascinating to behold. The wind being fair and a bunch of ducks conveniently placed for enticing, a wave of the hand sets the dog at our heels about his duty. He runs round screen No. 1, and hops over the first dog-jump. Immediately he comes within view of the birds, who, impelled by the curiosity of their kind, stop feeding, up go their heads, and a hundred amused and twinkling eyes are bent on the movements of the strange new creature that has broken in upon their repose. He disappears and re-appears again, his conspicuous yellow coat showing up well against the dull, winter-bare trees and the crimson twigged willow bushes, and presently with one accord the fowl are after him, streaming up the "pipe," their heads turning this way and that, right under the noses of the men watching, who must keep as still as mice until the last has gone up. This part of the business requires care, but if successfully managed, the keeper creeps down to the first screen and shows himself there to the birds in the tunnel, while he is still hidden from those in the lake. At once there is a clatter and splash, and, followed by the men, the birds hurry and scuttle up the tunnel, which narrows and contracts until the whole two or three dozen birds, it may be, are crowded in the pouch at the far end, whence they only emerge to be transferred, dead, to the ready sack.

Such is an exciting scene while it lasts, and more difficult to bring to a good issue in practice than it looks upon paper. In managing a "coy," so much depends upon keeping the pond at the flight season absolutely secluded and quiet. Anything will get it a bad name with the wildfowl, while, like Cæsar's wife, it should be above suspicion. Prowling gipsies, or tramps, spoil it for ten days at a time. The shadow of a hawk, a heron, or a fox, puts the timid mallards on the wing and sends them elsewhere. Even pike in the waters are objectionable; they have a decided taste for teal and young birds, and though the bulk and strength of a wild drake is proof against any such attacks, yet it is disturbing to its equanimity to see his smaller relations struggle, and splash and cry out, and then disappear stern foremost. The keeper, too, must know all about the right winds and weather, and something of the curious and punctual habits of the birds, whence they come, and when they are to be expected. Pochards, for instance, he will never try to capture in his long tunnel, for they invariably rise and fly back when alarmed, and a few birds escaping like this will spread the news. He must be clever in the feeding and management of the tame decoy birds, which by swimming about at all times in the mouth of the drains, bring the wild ones down as they pass overhead during their migration, and also unremitting in his guardianship of the place, and ready to turn out at two or three o'clock, it may be, in the cold winter mornings when the flight is on, to clear the channels, and break up ice formed round them. Perhaps the trouble attending their proper upkeep, the modern scarcity of ducks in paying numbers since the fens and moorlands have been drained, or a change of fashion, is responsible for the decrease in numbers of the ponds formed for this method of taking ducks. Probably in all the eastern counties there are not more than four or five actually working decoys, and Sir Ralph Payne-Gallwey, in his "Wildfowler in Ireland," says he only knows of three working in that country, viz. Mr. Longfield's, at Longueville; Lord Desart's, in Kilkenny; and Mr. Webber's, at Athy.

As to their origin, it is difficult to speak for certain. Camden says that 3000 ducks were sometimes driven into a single net at once. Willoughby also, speaking of Deeping Fen, declares that as many as 400 boats were employed, and that 4000 mallards have been taken in one driving. All this seems to point to the practice of driving young or moulting birds into a funnel-shaped net, somewhat like a modern decoy—a practice formerly carried to such an excess that an Act of Parliament had to be passed to suppress it. Spelman says, that Sir William Woodhouse, who lived in the reign of James I., made amongst us (*primum apud nos institutit*) the first decoy for ducks (*decipulum anatarium*), called by the foreign name of "a koye," apparently introducing a new word; and although he may not have actually been the first to take ducks by means of nets artificially arranged for that purpose in the form of a modern decoy, it is highly probable that he did introduce some important improvements, possibly the use of decoy ducks and dogs, of both of which he speaks. Mr. Thomas Wise, in "A History of Paganism in Caledonia," tells us that decoy birds for taking ducks were used by the most ancient tribes of the Pictish race, but he says nothing of the method.

The wild duck, in its many species, lends itself to the ingenious devices of many fowlers, who pursue and entrap it remorselessly, whether they be fur-wrapped Esquimaux on the Greenland Fjelds, wandering Tartars, mild but cunning Hindoos, gentle and persevering children of the Flowery Land, or, it is safe to say, the "sportsmen" of any other nation under the sun.

Speaking of the Chinese recalls one picturesque method they have of taking the beautiful painted teal of their woodland lakes. It is an aristocratic pastime, and requires specially prepared canals and embankments for its enjoyment. The gardens surrounding the palaces and great houses are always well watered by numbers of small streams, natural or artificial. Those which it is intended to devote to duck hunting are led by very tortuous courses through deep channels, with almost perpendicular sides, hither and thither amongst the mulberries and crimson-flowered rhododendrons of the extensive gardens. Ducks of several varieties—everywhere numerous in China—frequent these winding watercourses in considerable numbers, and when the mandarin or his high official determine on a teal catching expedition they go forth each armed with a thirty-foot bamboo, at the end of which is a stout and deep net. With these they cautiously approach the streams, and owing to the high banks are able to actually overlook the water before the ducks are aware of their presence. The astonished birds then spring up fast enough in a brilliantly coloured cloud; but the Celestials are ready for them, and as the gigantic "butterflies" top the grass and flowers the nets are brought into play, and half-a-dozen or more out of each school are enclosed and brought struggling to the ground.

The purpose of leading the streams in winding courses is in order that an attack on the ducks in one reach of water may not disturb those out of sight round the bend in the next. A curious scene it must be: the quaint and rich silk dresses of the men, bright sunshine on flowering shrubs, and the gay teal in their regal livery dodging the long nets—an admirable suggestion for a new series of "willow-pattern" plates.

Another method, slightly different in its earlier stages, but ending in the same way, has been mentioned by Mr. J. E. Harting. He says: "During the winter months many kinds of waterfowl resort to the inland pools which at the other seasons of the year keep to the sea or mouths of rivers. On these pools the fowls are allured—by food and decoy ducks—into so-called pitfalls, covered with rushes or fine nets, on either side of which are posted the beaters and sportsmen. As soon as a sufficient number of ducks have been allured into these decoys, they are made to rise by a loud noise, and the sportsmen take them with a sort of strong butterfly net. Those that escape are pursued by the hawks."

This is all very well for those who look chiefly to amusement, but the professional wildfowler has to adopt more wholesale methods. He resorts to netting and poisoning; the latter method is applied as follows: Rice is steeped in a decoction of coculus indicus, and then exposed where numerous ducks, etc., are likely to come. The next day the dead bodies are collected and sent to market. I never heard of anybody suffering any ill effects from this plan, which seems rather hazardous. The net used for trapping is usually a large pulling-over one, like those in use by the bird-catchers in the London suburbs. As it is used a long distance from the shore, another has to be sunk at the spot to prevent the ducks getting away by diving. The shallow water of the lagoons, nearly always seven feet in depth, affords facilities for fixing the nets, and live decoys are pegged down round them; the pull is up to three hundred yards in length, and is worked from boats.

Not only in the land of pigtails, but all over Asia, the duck tribe migrate at various seasons, and are taken in thousands. Mr. W. W. MacNair, who went in disguise through Kafiristan, a country between the Hindu Kush and Kunar ranges on the north-eastern side of Afghanistan, as yet sealed to Europeans, speaking of the Bogosta valley, says: "Between Daroshp and Gobor I noticed several detached oval ponds, evidently artificial, which I was told were constructed for catching wild geese and ducks during their annual flight to India, just before the winter sets in, about the middle of October. The plan adopted, though rude, is unique in its way, and is this. By the aid of narrow dug trenches, water from the running stream is let into the ponds and turned off when full; the pond is surrounded by a stone wall high enough to allow a man, when crouching, to be unobserved; over and across one-half or less of this pond a rough trellis work of thin willow branches is put up; the birds on alighting are gradually driven under this canopy, and a sudden rush is made by those on the watch. Hundreds in this manner are daily caught during the season. The flesh is eaten, and from the down on their breasts coarse overcoats and gloves are made, known as *margaloon*."

Again, on the lakes in the Cabul highlands, in which, during the rains, these birds abound, the natives adopt another ingenious plan for their capture. A small hut, covered with reeds and boughs of trees, is erected over a water channel that leads off the water into the adjacent country. After dark, when the ducks are floating about in the careless security of sleep, the trappers enter the hut, and opening a sluice gate, strike a light inside their watch-tower, and await the arrival of the ducks, which are soon carried by the newly produced current into the channel over which the hut is built. They enter through a narrow opening, and are seized with ready hands, and made lawful food by having their throats cut. In this manner a couple of men can easily secure from one hundred and fifty to two hundred ducks in a single night.

Not more than a month or two ago, the *Sporting and Dramatic News* had some sketches showing a common Indian trapper's dodge. They prepare a number of calabashes, from rind of the melon or gourd, and keep them floating up and down the lakes, on which swarm innumerable quantities of wild duck. From habit, the birds soon come to take no notice of the calabashes. The Indian, observing this, then prepares a calabash in which he cuts holes for seeing and breathing, and places it over his head. With this, and a belt round his waist, he starts on his duck-catching expedition. He is almost as used to the water as the prey he is in quest of, easily stealing quietly towards the flock, and when within an arm's length of a duck catching it by the legs, and before it has time to utter a solitary "quack" he whips it under the surface, and hangs it to the belt, very speedily filled in this manner. Our journal, unless I am mistaken, mentioned this practice as being in vogue in Yorkshire; and in fact it is curiously widespread.

On one part of the American coast there is a similar expedient practised, only that in this instance the headpiece is a cap of rushes—a number of them being always left floating about on the surface of the water to accustom the fowl to the objects, otherwise the process of capture is just the same as that detailed above.

It is also known in China, and Mr. Thomas Wise asserts,

in "A History of Paganism in Scotland," its use in the ancient Pictish Kingdom.

Across the water, in France, it was stated lately, in the *Shooting Times*, a clever and artistic method of taking black duck is practised, which might be adopted with success in other regions besides the neighbourhood of Cape Griz-nez, its chief home. The quarry is captured in this manner: At low water, or very near it, the fishermen, who chiefly use this method during their enforced inactivity in winter, go down to the sandy flats, and there selecting one of those beds of shell fish which must be familiar to all who have any experience of shore shooting, they drive a number of stakes into the mud, each stake being about three feet long and standing clear of the "flats," about two feet. To the tops of these, which stand in an oval shape and parallel to the coast, is stretched a long, large, fine-meshed net, as tight as it will go, and bound to the stakes by cords. To seaward of this a narrow, upright wall of net is also fixed in a crescent shape, its purpose merely being to act as a stop net, and to prevent the floating out to deep water of any dead ducks which may come loose from the main net. The latter, it will be noted, is stretched horizontally over a considerable space of the birds' choicest feeding ground. Matters having been thus arranged, the men return to their huts. While they smoke and amuse themselves the tide comes in, and with it come the black duck eager for food, and diving continually as the shore is neared. Little by little they approach the fatal spot, and the water now being two feet above the snares no harm is dreamt of. They swim and dive this way and that till at last the toil is under them. The leader has perhaps brought up a delicate morsel from the very limit of safety outside the net, and swallows it on the surface before his admiring companions. He prepares for another dive, but now the tide has drifted him over the meshes. Down goes his head, and with a whisk the tail disappears. He plunges under, and in less time than it takes to write is held firmly below by the strings into which he has thrust his neck. His companions note the prolonged dive, and, probably thinking he is having an especially good time of it, follow him, head after head being driven through the small but elastic meshes of the net, whence there is no return; and the unfortunate birds are held thus until they are drowned. Any which wash out as the tide recedes are caught by the crescent-like wall whose top is only just below high-water mark, or the fishermen come down to the beach with their poodles, and send them in after any ducks which may be floating away to sea. In this manner considerable numbers of birds are taken; but the profit is small, the victims selling for as little as fivepence apiece on account of their fishy taste and rankness. The black duck, it may be remarked, is the only form of flesh allowed to be eaten on fast days by the See of Rome, a curious bit of Pontifical irony, since this single exception is of a kind too rank to be touched by any but the very poor.

Sometimes ponds and lakes patronized by water-fowl will be unapproachable to the shooter for want of cover; he may nevertheless be able to obtain a few brace by one of the following methods. Let him take some good strong rabbit traps, and pour melted pitch on the plates. Before the pitch has time to cool, sprinkle on it several grains of barley. Choose a moonlight night for the experiment, and hang the traps, duly set on the side of the pond (within a few inches of the water) opposite the moon, so that her rays fall well on the pitched plates, which will glitter, and render the barley clearly visible to the ducks as they swim about the pond. Hang the traps on short pegs, so that when one of them is sprung by a duck "bibbling" against the barley, it may fall into the water, carrying the unfortunate drake with it; and if the trap be a heavy one, and the water deep enough, there will be little or no spluttering to alarm the other birds. I have never tried this plan, and therefore cannot speak personally as to its efficacy, but an old boatman assured me he had often done it successfully.

That boatman was a poacher, whatever he may have thought of himself; and perhaps it may be as well after this instance of treacherous ingenuity, to turn to an honester theme and outline a rough day's sport on the Scotch border fells and sea-shore, looking for our game honestly, and bringing it to bay with "straight powder" and in open day-light, in all the "pride and circumstance" of straightforward sports-craft!

## WINTER SHOOTING IN THE HIGHLANDS.

"Eight o'clock, sir!" says my faithful henchman, coming into my room with the hot water, adding, in answer to my sleepy inquiries, that "it's a fine morning, but freezing hard." Of the latter fact I have an instinctive perception in spite of the snugness of my retreat; that sort of feeling which warns one how unpleasant it will be to get up when the operation becomes absolutely necessary and can be put off no longer. On this occasion the subject seemed to require special consideration, the *pros* and *cons* of immediate rising being weighed with much deliberation. To begin with, the advantage of staying where I was appeared too obvious for a doubt. On the other hand, the first gong had sounded twenty minutes ago, so breakfast must be ready; possibly my hostess was already down, and, assisted by her three delightful daughters, presiding behind the silvery bulwarks of steaming coffee-pots and urns. I even fancied I could catch a faint whiff of all sorts of good provender on its way from the kitchen regions, and this fact was conclusive. Without venturing to think more on the subject, I muttered a once, twice, and away, and found myself safely standing on the floor. To draw up the blinds was the first operation, and there lay as wonderful a stretch of ice-bound country as any I have ever come across. The wild highlands of the western Scottish coast, and such it was that lay before me, are one thing in the summer, but quite another in the winter. To most they are only known when the land swarms with tourists, when every shooting lodge is occupied to overflowing from kitchen to garret, and gay picnic parties hold high frolic in each glen far and near. At that time the country is knee-deep in purple heather, the guns of the shooters are echoed on every side, and the grouse, doubtless cursing the inundation of sportsmen with modern fashions, long once more for the comparative peace enjoyed by their primogenitor, who had nothing to fear but his natural foes the hawks and the flintlocks of the highland chief's foresters. Every brook and tarn in June is threshed by lines of enthusiastic fishers; the post comes twice a day; smart equipages imported from the Lowlands dash about the country roads; and Scotland then is popular, wealthy, and overrun. Nearly all in these days of cheap tours know this phase of the matter, but when the first frost takes the colour out of the heather-bells, and the rowan-berries are at their brightest scarlet, a great change comes upon the face of the land. At the first pelting hailstorm from the northward darkening the faces of the lochs and filling the higher mountain gulleys with whiteness, the fine-weather invaders take the hint, the lodges are deserted, peers and commoners flit southward, Government itself makes note of the altered circumstances, and posts are reduced to one per day or less, hotels close their hospitable doors, and all the land sinks into repose, the scattered permanent inhabitants and many-ancestored lairds, with patriotism enough to stick by their acres all the year round, waking one day to find themselves alone and winter palpably upon them.

Such, but briefer, as befitted the coldness of my position before the window-panes, were my meditations while contemplating a wide stretch of snowy hills on the first morning of a midwinter visit to an old Scotch mansion, a visit to be varied by some rough sport and skating if the frost held.

However, it won't do to keep breakfast waiting any

longer, so down I go, and am soon seated at a table decked with snowy napery and crowded with savoury comforts for hungry men, very welcome in such weather as this. At the head presides the hostess, and on either side are her three daughters, all expert riders and skaters, each capable of fishing two miles of river in good fashion, or bringing down their brace of grouse, "when papa shoots the moor alone," and yet possessing all those gentle graces that are the boast of their unmatched countrywomen. The laird comes in directly. He has been out to see his thermometers, of which three or four stand at various points of vantage, and rubs his hands and seems highly delighted as he reports fourteen degrees of frost during the night, an announcement which elicits much applause, as of course we are all keen "curlers" here, and our hopes of a good season for that ancient game have been rising higher and higher lately. Yet neither curling nor skating were our ambitions on this particular day, which was to be devoted to a raid upon numerous flocks of wildfowl that the cold weather had driven to a chain of neighbouring lochs and a marshy estuary through which the river emptying them ran into a land-surrounded arm of the sea.

Breakfast over, there was soon plenty of bustle in the gun-room, where a sturdy Gael was busy filling cartridge-cases and slinging guns to their straps. In rough shooting of this sort, and more particularly in cold weather, a gun that cannot be hung over the shoulder when there is no chance of a shot, is anything but a pleasant companion. Then an emissary from the kitchen regions appeared with cook's compliments and a suggestive luncheon-basket. This Donald shouldered, together with a bundle of wraps, and, taking our own guns and cartridge-bags, the laird and myself waved a farewell to the bright group in the porch, and marched down the drive to where a dogcart was in waiting outside the big gates.

What a happy experience a fine winter's day is to those

blessed with well-strung nerves and a healthy appreciation of the beautiful! A comfortable breakfast and a mild cigar glowing with seductive warmth under the observer's nose are important concomitants for due enjoyment of the scene! For my part, fresh from the tropics, in whose gorgeousness familiarity has bred a certain distrust, a snowy landscape and a frosty morning are full of quiet charms. The feet make no noise upon the soft carpet of snow, which, as dry as the sand of the desert, falls like dust from the shoes at every step, and goes flying in minature siroccos across the open plains of the lawns and carriage drives, piling itself up against the trunks of trees and roots of shrubs, and scooping hollows to leeward of them, just as the fresh northern air drives it. The boughs of the evergreens are loaded down to the ground with their white burdens, and if by chance a blackbird, scared from his feast of yew-berries by approaching figures, breaks away with a resounding chuckle, he causes a whole avalanche of glittering crystals to fall from the shaken boughs behind him. But in general everything is very silent; the birds are too much occupied in searching for food even to sing if they had a cause, and in the farmyards the sheep and kine stand knee-deep in snow and straw, their whole attention taken up with the fragrant hay being liberally dealt out by that leather-legginged shepherd, who stops his work for a moment to touch his cap as the master and his guest pass. Truly the cold, white reign of winter is not without a sweetness of its own!

A sharp spin of a couple of miles brought us in sight of a boathouse nestling amongst birches at the head of a long streak of pale water. The loch was shut in by high hills on one side and stretches of flatter ground on the other, more level only by comparison, for it was marsh and bog plentifully supplied with deep peat holes and crevices broad enough to swallow a Highland cow, like the giant in the fairy story, "horns and all." Strange things are found in these steep-sided cavities. I have myself rescued from one such trap an imprisoned sheep suffering the last stages of exhaustion and starvation, while a curious story exists of a brood of half-grown flappers having been found in another, which they had entered along with their mother when very small, and, not possessing her powers of flight, had been unable to leave it; a little water in one corner and a few casual insects, we must suppose, supporting life in this novel open-air pen. For this region of dyke and pit we were soon embarked in a regular Highland skiff, impelled by the keeper's sturdy arms (the gillie who cannot row and doesn't look upon the water as a legitimate part of his territory is of little use on this side of the country); ten minutes and the peat banks of the opposite shore are over our prow, the bare wiry stems of the heather making tracery against the sky and looking like cotton plants in pod, with their weight of snow and rime. Donald shoves our bows between two rocks and deftly scrambles ashore with the rope to make it fast; but almost immediately crouches down, and we hear the mellow quack of a mallard which rises through the air from a pool within easy shot, but goes away unhurt, as, of course, we are not loaded. This quickens our expectations of sport, and we are soon landed, collars up, guns under arms, and ready for the march.

A snipe is the first bird to fall to the laird's gun, another getting up to the shot for me and dropping to the right-hand barrel. This is decidedly cheering, and we plod along enthusiastically over the crisp herbage, the dog sniffing about ahead, but being rather heavily handicapped by the stiff going for a time until we reach better ground. Some of the long-bills rise wild at a couple of hundred yards or more from us and sweep away to the southward like brown leaves in a gale, picking up as they go others of their species, and this irritates my companion, who scolds "Snap" for what is not his fault; but we get chances now and again which throw a rosier light over the proceedings.

An hour's trudge brings us to the foot of the first sheet

of water, with four and a half brace of snipe to our credit. There we find Donald again reposing against a rock, the smoke ascending in ripples from his pipe, and the boat quietly secured to a convenient alder at his feet. Together we walk down the opposite banks of the brook running to the next "lynn." Pleasant enough in the summer time, when its deep pools hold excellent trout, it now looks icy cold, and we wonder at the taste of a pair of water-ouzels, who stand on the stones bobbing their tails, or skim away down stream at our approach, in remaining faithful all the year round to such a desolate region. Nothing rewards us here until the far end is reached. At that spot is a bit of level ground, sometimes submerged by floods, and now a chequered surface of grassy "hassocks," surrounded by patches of ice and snow. No sooner do we turn the flank of a protecting spur and come upon this favoured region, all beglittered in the sunlight with icicles and frost, than a flock of teal spring from their cover and wheel into the air in front. H——, whose motto for to-day is certainly "ready, ay, ready," takes them "on the hop," and grasses one in good style. My first chance is at a "skyer," who doubles up and comes down back foremost forty yards distant, and my second barrel wings another lightly. We pick up the slain, their beautiful plumage contrasting wonderfully with the snow on which they lie, and then the dog goes for the wounded bird, recovering it after a chase over crackling ice, hardly stout enough to bear a mouse's weight, which lets him into some coldish water, if we may judge by the vigorous shake he gives himself subsequently. There is, to me, no water-bird like the teal for game qualities; he has "all the instincts of a gentleman;" powerful on the wing and sharp in his rise, he is up and away with half the fuss of any other duck, yet a light touch stops him, and unhit he often has the consideration to come round again after a shot if the sportsman keeps quiet. This latter quality was not illustrated by our teal to-day, so we beat down the water, disturbing some widgeon which could not be reached, and picking up three more snipe from a bed of reeds, a moor-hen, and a couple of wild ducks, all of which trophies took their way to the sad republic of the gamebag consecutively.

And then we lunched; the short winter day of high latitudes almost spent, and a choice bit of ground for "cock" yet to be searched. We took our meal under the lichened shelter of some birches, weather-beaten and dwarfed by repeated gales blowing down the neighbouring corrie. At our feet sparkled a fire of pine branches drawn from a dry corner under that rock which served us as a comfortable seat and table when a cushion from the trap that had brought along our provender was placed across it. The cold game pie was both juicy and tender; the "October brew" from a stone jug was amber clear, and as sparkling as Moët's best, and an inch of ripe and crumbling Stilton with a "short" sip of Glenlivet put the finishing touches to the sufficient if frugal refreshment.

It took us about as long as our cigars lasted to follow the smooth course of a roadway up a ridge, across its brow, and down the opposite glacis. From the top we saw the wide plain of the "mournful and misty Atlantic" looking black as ink amongst the framing of snowy hills on every side, but under us the warmer shelter of sloping plantations of larch and holly, cut up with water channels and dotted everywhere by dark towering heads of pines and strong young spruces.

There was little time to spare, so a couple of spaniels that arrived in charge of a boy from the keeper's cottage hard by were turned in, and soon the ball was going merrily again as they quartered the cover scientifically, and we walked silently behind down the parallel spinneys. The rabbits alone were numerous enough to have employed half a dozen guns, and flashed hither and thither in tempting style, a dozen or two paying the penalty of their rashness.

As for the woodcock, on whose behalf the expedition had been undertaken, there were not enough guns to do them justice. We wanted some outside the copse to interview *Scolopax rusticula* as he flitted from one shelter to another; but still we got an occasional glimpse at a retiring form clad in autumn russet, and in the majority of cases, if the chance was anything like fair, the bird was accounted for with little delay. A lordly cock pheasant rose near the laird, and was skilfully grassed by him ere the noisy bird had topped the neighbouring oak trees. Directly after this I managed to stop off my left shoulder a hare which was apparently starting for a journey to the other end of the kingdom, just as I was in the agonies of struggling through a holly hedge.

This lent variety to the bag, and was the last shot of a pleasant, if not very productive, day. We walked to the lodge, whose gates opened upon the high road, and, having warmed ourselves at the gallant blaze burning in the open hearth, were about to mount the dogcart for home, when there came the sound of bells outside, and a minute after in rushed Miss Mary. "Oh, papa!" she said to the laird, "you must forgive me for coming without asking you, but it is going to be such a beautiful night, and Madge and I couldn't resist the temptation of bringing the sledge for you instead of allowing you to drive home in the stupid old dogcart outside!"

The culprits were forgiven, and soon my entertainer was seated in front of a smart Canadian sledge, one of his daughters beside him, while I, having refused to take the reins, occupied a back seat with the other young lady, an arrangement much to my satisfaction, since I was allowed to light a meerschaum and keep my hands under cover of the heavy fur rug.

Sardanapalus offered half a year's revenue for a new pleasure! Did he ever try sleighing on a moonlight night? It is most delightful and novel. Not a sound broke the stillness as we sped along but the thin tinkle of silver bells on the leader's harness (for we drove tandem), he sniffing the fresh, cold air, and tossing about his head in wonder at the unusual pathway. Our runners passed over the dry surface of frozen snow with perhaps the faintest of murmurs, such as the ripples of a tideway make against the sides of a motionless vessel, but all else was hushed. At times we were floating down narrow gulleys between overhanging rocks where a streamlet, too lively to freeze, ran by the road-side, its course overreached with white crystals, and meandering through caverns and wonderful palaces of icicles and frosted herbage. All around nature was shrouded in white, on which the brilliant moon shone, and some of the bigger stars twinkled with unusual lustre in the deep blue vault of the sky. Again we would approach the outskirts of a vast pine forest, and, plunging in, leave the light behind, taking our way along with a strange association of speed and silence until we could almost fancy we were disembodied and going to some Walpurgis revels! "Do you think there are *any* wolves left in England now?" inquires my companion in a hushed voice, glancing round at the sombre aisles of the dimly seen woods, where disjointed fragments of old mountains take strange forms as rays of moonlight steal down here and there to light them.

I assure her there is nothing more wolfy in the neighbourhood than the skins of a couple of those animals forming the rug that wraps us both, but she is very silent until we pass into the moonlight again. Then comes the run home along the other side of the valley, the lights of the hall twinkling out in the darkness; the arrival and confiding of the steaming horses to the ready stable-boys, and we peel off our furs and wraps to follow the genial old laird into the dining-room, where he forthwith concocts with due solemnity a brew of hot punch in an ancient wassail-bowl, of which we all taste, and so for the fragrant "half-pipe," and to well-earned rest.

# CHAPTER IX.

# *SEA FOWL.*

## FRIENDS OR FOES.

HAS the Sea Birds' Preservation Act failed by over successfulness or by under; are we unduly protecting the gulls and guillemots to the ruin of our coast fisheries; or are we negligent and insensible to the exterminating ravages of cockney sportsmen and plumesters? Such questions as these are frequently asked and answered with every variety of conviction and logic. My own opinion, I may say at once, is that over preservation of the bird life of the sea-shore and marsh flats is simply and utterly impossible. If we were to infence our seafowl with legislative protection, until they were as common as sparrows in a winter stackyard, I do not believe the price of herrings or sprats would go up a farthing a "last" from this cause. That thousands of fish might daily go down these myriad hungry maws is quite certain; but against this there is the fact, never sufficiently recognized, that in the economy of such things as the herring shoals, it is space and opportunity alone which limit their reproduction and increase. The onslaught of a hundred thousand solan geese and puffins could be repaired by the fertility of a few score female herrings, if Nature found there was sea room and food sufficient for them. Of this we are as confident as that Providence understand such matters as well—if not better—than the town council of Little Pedlington-by-the-Sea.

I myself have a very certain admiration for the herring; the salmon may be the king of fish, and the pink-fleshed loch trout of Scotland make epicurean mouths water at the antipodes; the white fillets of sole may be more aristocratic, and the creaminess of a seasonable turbet unique, but none of these have anything like the savour of the necessary, harmless bloater! He is at the bottom of the scale in humility and consideration, yet surely a long way from the last in all the qualities that could endear him to the hungry and frugal. If it was a case of kittiwakes or red-herring, then patriotism, as well as the remembrance of a score of simple meals in quiet hostelries, and the snug parlours of water-side inns, would cast judgment in favour of the latter. But matters have not come to this pass; the world is quite big enough for fish and feathers, and this in spite of an avaricious commerce or the mercantile greed of some few long shoresmen, who take an undoubtedly heavy toll of the harvest of the sea.

That the seafowl do a scarcely appreciable amount of harm from a utilitarian point of view is not difficult to demonstrate to an open mind. The chief culprits accused of voracious and misdirected appetites are the common gull, black-headed gull, herring gull, great black-backed gull, cormorant, green cormorant, gannet or solan goose, guillemot, puffin, razor-bill, northern diver.

Besides these there are some culprits in a lesser degree, or whose interference is so occasional as to be hardly worth considering. There are rarer gulls than the five mentioned that now and then mix with the flights and feed amongst them; the ducks of a dozen species are also omitted, as, though many of them are at sea all day, they are vegetable feeders. The same applies to geese and swans; and godwits, sandpipers, and plovers, are harmless dabblers in back waters and creeks, where they thin out the small crustacea and shrimps.

The main charge against all these birds is, of course,

that of diminishing national supplies of food by pillaging the herring shoals and schools of edible fish. It must be remembered, however, that gulls, at all events, are no divers, and the herring usually lie a fathom or so under the surface. A kittiwake, or "cobb," has to take what he can glean on the surface; he will swoop round and round a turn or two in the sky and drop down with astonishing precision and exactness on anything he cares to pick up, but he does not go under, and rides in the hollows of the waves as lightly as a cork. His food is flotsam and jetsam—the off-washings of the shore and all the disjecta of the sea bottoms, the soft shelled crabs that come to the top, the sickly or wounded fish, and occasionally some of the small fry the observant boatman will have noticed basking in the tepid water shining under a summer sun, or flashing into the air and daylight as some "ravening salt sea shark," some great bass or whiting of the weedy ledges, runs amuck through their close-packed columns and drives them up. Indeed, in helping themselves to the young of these same whiting, the teeming "haddies" of the Scotch estuaries, the gulls do immense service, for big fish are to little fish far worse foes than anything wearing feathers.

The Yorkshire cragsmen who live amongst the cliffs all the time the birds are breeding and have daily experience of their housekeeping arrangements, describe the fish remains littering the cliff-shelves as chiefly those of "base" fish, sand eels, gobbies, wrasse, and the like. Herrings, of course, in any condition were absent. "Those persons who write so glibly on the subject of the destruction of fish by sea birds," writes the Rev. F. O. Morris to the *Yorkshire Gazette*, "forget that long before guns were invented the birds must have had it all their own way on the cliffs of our coast all round these islands; and how was it then they did not exterminate the fish in ages long ago, instead of their increasing in the way they have done?" According to the Rev. Barnes Lawrence, it is "not so much a question of how much the gulls eat, or how many birds there are to eat the fish, but what fish they eat, and what other fish have a better chance in consequence."

The economy of Nature is a mosaic from which the absence of a single part loosens all the neighbouring structure. Were there no check upon the whiting and such other destructive fish, supplied by the gulls who feed amongst their young, then these might play havoc in turn with the herrings. Nor does this argument clash with that of the immense prolificness of food fishes, because man, demanding an undoubtedly heavy toll of good fish, and not paying an equivalent amount of attention to their enemies, these foes must in turn be kept in place by some means such as the predatory birds supply.

Mr. Morris, the well-known author of "A History of British Birds," a charming and invaluable work, has lately made some calculations regarding the harm which the wanton slaughter of sea birds effects, and though his deductions lay him open, I fear, like all such attempts, to hostile criticism, they are curious and interesting. Having summarized the number of gulls killed in a season along the Yorkshire coasts alone, he adds: "If we carry on our calculation still further, say, if each bird dives nine times per hour (I believe eleven is the usual number) and catches three whiting per hour, or one in three dives, we have:—

975 birds killed daily for "pleasure."<br>
109 ,, average for professional bird killers.<br>
———<br>
1,084 killed or wounded daily.<br>
3 whiting.<br>
———<br>
0,052 per hour.<br>
12 (say 12 hours per day diving for food).<br>
———<br>
39,024 whiting destroyed per day.<br>
110 days.<br>
———<br>
4,292,640 whiting destroyed in the breeding season.

Mackerel, herring, sprat, and haddock are more particularly regarded as "food fish," on which the young of whiting feed. And allowing each whiting to eat 200 "food fish" during the 110 days, or while the birds are with us, we find :—

| | |
|---|---|
| 4,292,640 | |
| 200 | |
| 98,528,000 | "food fish" lost by the destruction of birds in 110 days. |

This deduction of nearly *one hundred million herrings shot away with the lives of the kittiwakes and gulls every season,* under one line of cliffs alone, is a rough, unscientific perhaps, but nevertheless effective popular argument for the good cause, and should make the owners of the *Sarah Jane,* the *Two Brothers,* and every other North Sea yawlsman rub their chins reflectively and reconsider their ill-will towards the birds, or their willingness to show the gentlemen of the Sheffield furnaces and the Midland cotton mills the breeding-places of the fair white fowl that supply the life and pleasure of the great north seas.

Nor are the fishermen the only class who reap some benefit from these tenants of the crags. Gulls wander in-land, especially in stormy weather, and though never so omnipresent as rooks and starlings, nor so keen in the farmer's service, yet they do him some good work such as one of Mr. Morris's correspondents points out. He writes: "I am game watcher to Lord Londesborough, and have been for over twenty years in his lordship's service, and I have seen a good deal of destruction of sea-birds, and have lived in the neighbourhood the greater part of my life, and shall be very glad to give you all the information I can, respecting the destruction of sea-birds. I think it would be a very good thing to prolong the preservation from the 1st of August to the 1st of September, and I consider the month of August is the very worst month in the year for the destruction of sea-birds, for the greatest part of the young are helpless in that month. After there has been a party of shooters, the beach and the cliffs are strewed with young ones. I fell in with a party one day myself who had been shooting. They had caught four young guillemots alive, and the poor little things were yelping themselves to death all the way they went. They had got one kittiwake with a broken wing, and were carrying it with the other wing. I asked them what they were going to do with them, and they said they were going to take them home with them, and turn them into the garden. If that is not cruelty to sea-birds, I do not know what is. I think it is a very great shame to shoot gulls and kittiwakes at all, for they are the best friends the farmers have, for they never touch a grain of corn at any time of the season. I think I need not confine myself to the farmers only, but I might say the country at large, for all the trades are upholden by the farmers. Forty years ago we never had any grubbed land in this neighbourhood, when we had thousands more gulls and kittiwakes than we have now. They used to follow the plough by hundreds; the ploughboy could turn round with a stick and hit them; now he may plough for days, and never see one near at hand, and we have very little land in the neighbourhood but what is infested with grubs. There was a gentleman farmer in Buckton some years ago, who shot a gull, and he said he was fit to cry when he saw what a friend he had shot, for when it fell it threw up a quantity of nothing but grubs and worms, and he vowed on that day that he would never shoot another as long as he lived. Some people say that they are very destructive amongst fish, but I think what they get is a useless kind of fish, for what the climber has brought up to me are almost as much in the shape of a worm as a fish. I must admit that they will want a great quantity of food of some kind, as many of them never feed on the land; but forty or fifty years ago, when we had thousands more sea-birds than we have now, I have taken tons of fish from Bridlington to Hull at sixpence per stone."

Several species build on the inland moors and wastes, and

then the jealous eyes of the keeper sees first-class misdemeanants in them. One declares that a big nesting gull will quarter the hill-side for young game like a hen-harrier on the marsh lands. I must acknowledge in reply to this that if I were a young grouse poult, with a wiry hank of knotgrass by some mischance "clove hitched" round my leg—my comrades, too, over the brow of the hill—then the wide pinions and the keen brown eyes backed by the remorseless bill of a big gull would not be the sight I should best enjoy seeing to windward! But these gulls hunt the moor sides for mice, frogs, lizards, and so on; they keep chiefly to the parts of the heath which grouse and blackgame avoid, and I do not think a colony of them would do any serious mischief to a moor on which the game was healthy and not overcrowded,—the latter a condition of affairs which Nature abhors and takes the first means at hand to mend.

As for the rest of the list of sea fowl generally regarded with hostility by some folk or other, there are amongst them birds which undoubtedly sympathize with human fancies in the way of a fish diet. There are the divers—the "loons" of the boatmen, extraordinarily voracious and expert fishers; but then there will not be more than a pair of them to many miles of coast. The gannets, again, I fancy, appreciate "caller herrin" as much as any Loch Fyne housewife. It is truly a fine sight in free falconry to see that great white body of feathers and strength, a hungry solan, sweep down the rifts of the clouds, surveying as he goes the hollows of the waves that toss by under him in long confused ranks before a fresh off-shore breeze, and then mark him suddenly check his easy sweep from point to point and fall like a white satellite with a triumphant scream from just under the grey sky into those green waters which close over him in a cascade of white foam. If any one could take their eyes off the bay before he is up again, mounting in easy spirals to his watch towers in the rift—or begrudge him that silvery fish (whatever it be) over which the wind brings us his wild exulting laugh—then we can only suggest they are more conventional than we are and less easily pleased.

The green-eyed cormorants are familiar objects of the coast, either flapping with undeviating integrity of purpose just above the water across the harbour mouth, or "hanging themselves out to dry" on the warm rocks after a successful foray. A well-wisher of theirs puts in a kind word for them. "Nor from another standpoint can the cormorant be regarded as injurious. I do not refer to any qualities which might touch the heartstrings of the æsthetic or sentimental, which vibrate so plaintively for the captive goldfinch or the tender pigeon's wrongs, for this is only a black, ungainly fowl, albeit beloved by Njörd of Northern lore—a patient, clever fisher, but of what? Often I have watched the cormorant fill its pouch before taking its nine-mile heavy flight to its young on the cliffs of Budleigh Salterton, where, midway between the sea and heather, it breeds unmolested among grey, samphire-covered rocks, or ledges of red sand, and seen in nearly every instance its prey has been the flat fish or the eel, than which no greater enemy exists to salmon spawn and fry." And, further, what cormorant can compare in destructive capacity with the greedy fisherman, or poacher, who kills the salmon big with spawn for an uneasy meal or shameful market? In truth, the *Phalacocorax carbo*, as Temminck has it, has not alone the right to a name distinctive from the earliest days of rapacity and greed.

In Devonshire, we are informed, the responsible authorities silently proclaim their opinion of this great ungainly sea-crow by withholding protection from him all the year round. In China and Ceylon he is a professional fisher working from a boat's prow, with a strap round his neck, industriously and successfully. Except perhaps in the breeding season, when he, like all other animate life, has given hostages to fortune and increases his kind at his own imminent peril, the cormorant is very well able to take care of himself.

The sea birds have their protective legislation, and I am

not in any great fear of their speedy extermination. What, however, the Rev. F. O. Morris, and others equally perspicuous and kindly hearted, seek to do is to rouse and maintain a lively sympathy with our wonderfully rich and varied shore and inland fauna. They would forbid the cockney fusillades which sweep the English cliffs of their tenants *while the young are still callow* and dependent; nip in the bud, if I understand them aright, puerile and abortive superstitions regarding the misarraugement of Nature, and frown down (perhaps the hardest task of all) the shop-girl fancy for ill-gotten plumes —wantonly pillaged for a purpose they do not effect. These humanitarians, however, are no sentimentalists, or they would forfeit the support of the keen British relish for outdoor sports which vivifies and supplies the backbone of their cause. They recognize there is a difference between the barbaric carnage which loads the stem and stern of a boat with the shattered and soiled bodies of seamew and tern, of which little or nothing can be made, and reasonable and legitimate sport when the breeding season is over. It must not be forgotten that grouse moors and partridge manors are little less accessible to the majority of our countrymen than the golden fruit of the Hesperides. They turn naturally to the foreshore, that border country between riparian avarice on the one hand and the ocean on the other, and here it is only natural they should find some freedom. I myself have spent many happy days on the shingle and under the white face of the towering cliffs, matching my skill in stalking against the superabundant watchfulness of the curlews, or attempting to approach redshank and plover in wilderness of shingle and yellow sea poppies. To attempt the suppression of these proclivities in our race by Act of Parliament, would be as senseless as was the project of the emperor who sought to cure his subjects of avarice by coining money of preposterous weight and steeping it on the threshold of the royal mint in evil-smelling fluids.

But every true sportsman detests remorselessness, and

with the spread of good sense and the active propaganda of such kindly leaders as the rector of Nunburnholm, sea and land birds will receive due protection and recognition without, we think, the naturalist and gunsman's modest and orderly pleasures being infringed.

In the new edition of "The History of Foreign Birds," by Yarrel, the editor, a well-known member of the Zoological Society, writes thus:—

*Laridæ*, p. 653.—"The eggs are seldom laid until the last week in June, so that many of the young are still in the nest or barely fliers when the Sea Birds' Protection Act expires on the 1st of August. Some years ago, when the plumes of birds were much worn in ladies' hats—a fashion which any season may see revived—the barred wings of the young kittiwake were in great demand for this purpose, and vast numbers were slaughtered at their breeding haunts. At Clovelly, opposite Lundy Island, there was a regular staff for preparing the plumes, and fishing smacks, with extra boats and crews, used to commence their work of destruction at Lundy Island by daybreak on the 1st of August, continuing this proceeding for upwards of a fortnight. In many cases the wings were torn off the wounded birds before they were dead, the mangled victims being tossed back into the water. The editor has seen hundreds of young birds dead or dying of starvation in the nests. . . . It is well within the mark to say that at least nine thousand of these inoffensive birds were destroyed in a fortnight."

But those who like statistics of this kind ought to write to the Selborne Society for a useful little pamphlet published on the abuse of bird plumage as a means of adornment. We do not attach very much importance to figures, for we can judge for ourselves in the streets and shops of London, Paris, New York, and other large cities and towns, what must be the sacrifice of bird life; nevertheless we give a few items derived from various authentic sources. Between December, 1884, and April, 1885, there were sold in one

London auction room 6228 birds of paradise, 4974 Impeyan pheasants, 770 Argus (Monal), 404,464 West Indian and Brazil birds, 356,389 East Indian birds, besides kingfishers, parrots, bronze doves, fruit-eating pigeons, jays, rollers, regent birds, tanagers, creepers, chats, black partridges, golden orioles, pheasants, etc.; and various odds and ends such as ducks' heads, toucans' breasts, and sundry nests. "Wanted, 1000 dozen seagulls" (Advertisement, *Cork Constitution*). "Wanted, 10,000 pairs jays', starlings', and other wings." From America, we get the following. A Broadway dealer says, "We buy from 500,000 to 1,000,000 small American birds every year. Native birds are very cheap." Concerning terns, Mr. Dutcher says, "3000 were killed at Seaford, L.I., and 40,000 at Cape Cod in one season." One taxidermist prepares 30,000 skins for hats and bonnets every season. Maryland sent 50,000 birds, many being Baltimore orioles, to Paris in a single season; a New York taxidermist contracts for 300 skins a day, for his trade with France; Ohio Valley, 5000 skins. We might add pages of such facts. It is rather the fashion in England to say that these American figures are of no interest. But most of the birds are killed in America in a great measure for export to England, and thus the destruction of bird life is kept up by English women. Existence, to the Baltimore oriole and our robin redbreast, is equally enjoyable, Why cut it short? A bird-skin stuffed, wired, and supplied with eyes, lasts for a few weeks and is then throw aside as "out of fashion."

Do not injure the cause of the preservation of birds, Mr. George Musgrave advises, "by trying to prove too much, and in some instances appearing to value the lives of dumb animals above those of men and their families who produce or obtain food for the community." Sea birds have their faults. The skua bullies the gull, and the gull behaves infamously to the guillemot. The puffins evict the rabbit, and thus deprive human beings of food and a source of income. The mariner who trusts to sea birds in a fog or a storm (*pace*, Mr. Morris !), where they are very much at sea themselves, will never, we hope, obtain the command of an emigrant ship! And finally, in all friendship to Mr. Musgrave and his allies, I would suggest that not only is it judicious not to attempt to prove too much, but also there is wisdom and reason in not demanding too much. The poor shooter justly claims as much moral right to carry his gun under the cliffs in the hot autumn weather, as any virtuous friend of the birds may do to relish his tender spring chicken and bread sauce, or to take another slice from that confiding Michaelmas goose who put his trust in the motherly kindness of the henwife.

## CHAPTER X.

# *QUILLS AND FEATHERS.*

### SOME NOTES ON BIRD BOOKS.

THERE would scarcely be a better exercise for any one who might be inclined to doubt the abiding popularity of matters of ornithology and sport with the British public, than to take a short expedition into the literature of the subject. This has accumulated and still accumulates in a manner that is very gratifying to those who love the country side, but sorely perplexing to the assimilator who would reduce the chaotic mass of information into some reasonable form and order. To index everything that has been written upon ornithology for even the last hundred years would be to compile a vast catalogue, reaching the dignity of a portly encyclopædia, and to own all these various works in every written tongue, were it possible, would be to possess a magnificent but overwhelming library.

One thing simplifies the problem, and this is that the best works on this subject are without question amongst the most modern. There are no classics in ornithology. The occasional allusions in remote writers to the subject are often gems of description extraordinarily pithy and pointed because they came from direct, unprejudiced observation. What, for instance, could be more fascinatingly real than Virgil's account of a rock dove breaking from her cavern nest?

> "Qualis spelunca subito commota Columba,
> Cui domus," etc.

which Dryden translates with half the ring of the original—

> "As when the dove her rocky hold forsakes
> Roused in a fright her sounding wings she shakes;
> The cavern rings with clattering: out she flies
> And leaves her callow care, and cleaves the skies;
> At first she flutters—but at length she springs
> To smoother flight, and shoots upon her wings."

But such are incidental to other matter. Amongst the books on English birds which figure conspicuously on the naturalist's shelves and are dear to his leisure hours are such, for instance, as Yarrel's "**History of British Birds,**" with upwards of 1070 engravings on wood—containing accurate figures, with accompanying description of every known variety of British bird; and this has from the first taken its position as the standard authority on the subject in our language. Yarrel has been edited by Richardson, Newman, and others, and not neglected by the publishers.

Bewick's "**History of British Land and Water Birds**" is almost more famous for its woodcuts, full of animation and a quaint delicacy, than for its letterpress.

These volumes, in their many reprints and with their supplements, belong, we must confess, rather to the province of the bibliophile than to the ornithologist. Of the many issues, that of Newcastle, bearing date 1826, was the first edition in which the "Supplement" was incorporated, and also the last edition which the author-artist saw through the press. The paper on which this edition was printed is reputed to show the delicacies of the engravings to the best advantage. But, great as is our respect for this limner, he must be put down rather as an engraver than as a naturalist.

Then there is Sir William Jardine's "**Naturalist's Library,**" a bold attempt at summarizing Nature in forty volumes, more suited to the taste of the first half of the century than to this latter part. Sir W. Jardine's coadjutors in this admirable series were Swainson, Selby, Macgillivray, Waterhouse, Duncan, Hamilton, Smith, and others. There are some 1200 beautifully coloured plates in the work, a copy of which is perhaps worth five or six guineas.

Latham's "**General History of Birds,**" with the synonyms of preceding writers, and 194 carefully coloured plates, 11 vols. 4to, and printed at Winchester in 1821–28, is a well-known work. "If the author had used a more modern system of classification instead of adhering to that of Linnæus, this work would unquestionably be one of the most complete and useful in existence," wrote a contemporary reviewer. Considering, however, that the author was nearly ninety when his work appeared, it deserves much admiration.

Montagu's "**Ornithological Dictionary, or Alphabetical Synopsis of British Birds,**" with coloured frontispiece and 24 plates, in two volumes, dated 1802, is often quoted. Colonel Montagu was one of the few soldiers who devoted themselves to ornithology against a whole array of the church militant.

Gosse is a familiar name again. His "**Popular History of British Ornithology,**" a familiar and technical description of the birds of the British Isles, 19 plates, containing 70 coloured figures of birds (1853), is very pleasant reading. He has written, too, some "**Naturalist's Rambles on the Devonshire Coast,**" which are illustrated with coloured plates, and come near to the freshness of Gilbert White himself.

That latter admirable divine must not be overlooked. To say there is an indescribable freshness about his work, like the inalienable cadence which hangs round Shakespeare's sentences or the mellow vigour of Scott's prose, would be trite and ineffective. He is amongst birds what Isaak Walton was amongst fishes—the professor of the field, and the permanent holder of that chair which Nature herself has endowed.

Well known to every one for the delightful details it contains of the habits and manners of British birds, this work is interspersed occasionally with notices of other animals, but the amiable author appears to have paid most attention to the feathered tribes. The "**Natural History of Selborne**" has passed through a great many editions; Rennie's contains notes by Herbert, Sweet, Rennie, and Mitford, and should be in the hands of every one—the general reader no less than the professed naturalist. All scientific detail is here avoided, and indeed White probably knew very few of the Linnæan names, as we frequently meet with such appellations as "*Passer arundinaceus*," "*Regulus non cristatus*," etc. The book consists of a series of letters addressed to Pennant and Daines Barrington.

Then there is Thomas Pennant, the first three volumes of whose "**British Zoology**" can hardly be spared from our shelves, though the arrangement (of 1781) is rather out of date to-day. Side by side with him are Buffon's works, and the pleasant chapters of Wilson and Waterton. The latter was almost the first amongst naturalists to place the study of birds in their native state before their arrangements in cabinets and museum shelves. He invented a system of taxidermy which, like some ancient Egyptian arts, became extinct with its inventor; but any one who would know what a happy valley of bird life may be formed, even in this northern climate, should read the account of his English home and the wonders he performed there in taming and acclimatizing.

Macgillivray prepared an excellent "**Manual of British Birds,**" and Selby's "**Illustrations of British Ornithology**" are often quoted. These were, at the time, the most masterly works, on the whole, that had appeared on the birds of Britain. The first edition was on the system of Temminck, with one or two improvements, as, for instance, the removing from the genus *Sylvia* of Latham the common and gold-crested wren. The descriptions of habits, nidification, etc., are sufficiently full for a systematic work, and always correct. The plates are all drawn and coloured from Nature, by the author. Every individual of the families *Falconidæ* and *Strigidæ* would make a perfect picture of itself, so beautifully and correctly are they executed. "Few of the others come up to these, and we are sorry to add that the talented author has entirely failed in the delineation of the *Sylviadæ* and *Fringillidæ.*" The figures of the falcon and owl families have certainly never been equalled—even by Gould and Audubon.

This, with one or two omissions, brings us down to some more modern writers; the J. G. Atkinson (dear to school-boys) whose "**British Birds, Eggs and Nests**" have been the key to lots of delightful half holidays amongst English lads, and whose little classics bring back happy hours when the "boys of an older growth" chance upon them amongst their heavier volumes. J. E. Harting's "**Handbook of British Birds**" shows the distribution of the resident and migratory birds in the British Islands, with an index to the records of the rarer species. "**The Ornithology of Shakespeare,**" critically examined, explained, and illustrated, is a useful work not attempted before; while in "**Our Summer Migrants,**" we have an account of the migratory birds which pass the summer in the British Islands, illustrated from designs by Thomas Bewick. For those who reside in the country and have the time and inclination to observe the habits of birds, this is a most entertaining volume. The habits have been noted and much information generally given about our summer migratory birds.

Without our Rev. F. O. Morris, of Nunburnholm, we should be lost indeed! His "**History of British Birds,**" in six volumes, with 365 finely coloured plates (£6 6*s*.), and published only some fifteen years ago, could hardly be better. In the smaller editions since issued, the letterpress is reproduced in its completeness, but the plates have been cut down to a woeful extent owing to the exigencies of binding, completely spoiling their artistic appearance, though not their usefulness, of course, for purposes of identification. To the beginner, anxious to possess a reliable work, and yet uncertain what it should be, I would certainly recommend Morris—the larger edition, if it can be afforded (and it is sometimes to be had cheaply second-hand) ; and if not, then the lesser one.

To Harrison Weir the ornithologist owes a debt of gratitude, and the services of the Rev. J. G. Wood in popularizing the science will not be forgotten. Mr. Smiles, in his "**Life of a Scotch Naturalist,**" has done a good deed in showing the enthusiasm is no expensive hobby, but one that can brighten and ennoble the humblest existence. To Mr. R. Jeffries we look for some delightful sketches of natural history and rural life, in a vein that has been too much neglected of late; and so on through more well-known names and deserving works than we can find space to mention.

These have all, so far, been the student writer, the naturalists of pen and scapula; but there are others—the naturalists of gun and pen, whose writings are at least as entertaining, and indeed, sometimes more valuable to the cause of sterling science than the manual of the savant whose happy hunting-ground is the labour of his predecessors, and who never saw half the birds he described unticketed or full of any sort of individuality but such as arsenical soap and wire can supply.

If, as we have seen, the monkish writers attempted a little occasional descriptive ornithology, it was not long after this that the first quaint attempts were made at directing the "fowler" in his art. Not perhaps the first, but still an early essay, is the "**Boke of St. Alban's,** containing treatises on hawking, hunting, and cote armour," printed in 1486 by Caxton.

There is a curious little book on "**Hunger's Prevention,**" by one Gurvas Markham, and some others such. But the handler of modern arms of precision does not become at home, or begin to "feel the bottom," until he gets amongst such books as Squire Osbaldiston's "**British Sportsman,**" a dictionary of recreation and amusement, with copper plates of hunting, coursing, and shooting. This bears date 1792, and, at a time when there were few such, must have been delightful reading indeed. Such miscellanies were then in vogue, as the "**Gentleman's Recreation,**" in four parts—viz. hunting, hawking, fowling, fishing; also the method of breeding and managing a hunting horse (1721); or the "**Sporting Review,**" a monthly chronicle of the turf, the chase, and rural sports in all their varieties, edited by "Craven," with numerous illustrations (some coloured) by Alken and others. This contains complete articles on racing, fishing, coursing, hunting, shooting, coaching, yachting, etc.

Some of these occasionally come to light in old book boxes, and the bibliographic ardour of the age fixes a value upon them above their worth. Daniel's "**Rural Sports,**" hunting, angling, shooting, fowling, etc., with numerous beautiful engravings by J. Scott, in four vols., roy. 8vo, and dated 1812, deserves mention as a successful example of the pleasant-penned lexicographer, who thought nothing of summarizing a dozen sports which nowadays would be relegated to as many individuals. He is appealed to less as a guide to-day, than as a historic sign-post in the annals of sporting; and any one who would know how game was shot or hunted, while the century was still in bud, takes down their Daniel, and rarely in vain. His contemporary, Thomas, wrote a "**Complete Sportsman's Companion,**" with descriptions of the various kinds of dogs, their breeding and rearing; also instructions for shooting grouse, pheasants, and snipe, illustrated with four pretty etchings of shooting scenes by Howitt (1820). Maxwell's "**Field Book of Sports and Pastimes of the United Kingdom**" is a volume full of every subject connected with games and sports, with numerous woodcuts.

These, however, are but stars of the second and fourth magnitude in the firmament of our library walls, compared to that brilliant luminary, Colonel Hawker. His "**Handbook for Young Sportsmen**" is a priceless volume, in spite of all that has been written since. Guns have changed and circumstances have altered, but this does not affect Colonel Hawker, who is still our reliance upon everything connected with waterside shooting especially. The art of the covert side was not quite so dear to him as the freer and more adventurous sport of the marsh land and estuary, a peculiarity he has shared with many another keen gunsman and good observer. This writer was an early disciple of large bore guns, and a thorough "all round" shooter, than whom there could scarcely be a pleasanter friend for the fireside or safer guide to the common sense of the tide way.

The "**Oakleigh Shooting Code**" (1836) is often referred to. It deals chiefly with red grouse, blackgame, and partridges; and "Craven's" (Captain J. W Carleton's) "**Recreations in Shooting**" (1846) is a handy volume, very prettily illustrated.

This epoch was fertile in writers of the kind. Who could possibly overlook or fail to be fascinated by St. John's (Charles) "**Tour in Sutherlandshire**," with extracts from the field books of a sportsman and naturalist (1849). "One of the most agreeable mixtures of observation, description, incident, and anecdote that we have met for many a day."

Colquhoun's "**Sporting Days in the Highlands**" deals with wildfowl shooting, deer stalking, etc.; his "**The Moor and the Loch**" contains practical hints on Highland sports, and notices of the habits of the different creatures of game and prey in mountainous districts of Scotland, with instructions in river, burn, and loch fishing (1841).

"**The Wildfowler**," by H. E. Folkard, is another delightful book for sea shooters, full of wise advice about duck shooting with gunning punts and shooting yachts; as also much about fowling in the fens and in foreign countries, rock fowling, and so on. The steel plate engravings to this volume are both delicate and carefully executed, and the chapters are annotated and stocked with an infinite variety of information. This is another of those books which every one should possess.

We have left ourselves but little space for the writers of to-day, who, however, are no doubt fully able for the most part to call attention to their own handiwork. "**The Badminton Library**" is an ambitious attempt to sweep the board and summarize every English sport in one of a series of volumes. It will never oust the fathers of the craft from their places on our shelves, clever as many of its writers undoubtedly are in their distinctive branches. "Tegetemier on Pheasants," and "Idstone" on shooting them, go hand-in-hand; "Stonehenge" (the genial editor of the *Field*) has written handbooks of amazing popularity; and "Wildfowler" (L. Clements) revived, for the time at least, the passion for marsh and rough shooting, which, if it ever becomes extinct, will do so the rather because there are no longer any suitable spots where it can be practised, than because the race of to-day lack hardihood or manliness for its successful pursuit.

In commencing this chapter I had before me a vast amount of rough material in the form of endless cuttings—the gleanings of many months' industrious reading of book lists,—as well as notes from the contents of my own shelves. But it soon became obvious that to utilize even the greater portion of all this crude knowledge would necessitate the preparation of yet another volume to the naturalist's library to accommodate it. So it was ruthlessly jettisoned; and it only remains to add a word regarding one or two useful books on foreign birds.

Of course some of the naturalist-authors mentioned in the beginning may be consulted with advantage for the bird life of distant countries. But few Englishmen have exceeded Gould in the versatility of his knowledge on this subject, or the magnificence of the works in which he embodied it. Messrs. Southeran announce an edition of his complete labours in twenty volumes, for which they ask the sum of £400 per copy!

"The works of Mr. Gould constitute a new epoch in the history of ornithology, from the boldness of the plan on which they were executed; the number of new species added to science, and of doubtful species cleared away from previous obscurity; the unadorned fidelity of the descriptions; and the exquisite accuracy of the plates, in which the utmost adherence to nature is united with that felicitous effect which stamps the artist, and proves that grace and truthfulness may meet together. Again, Mr. Gould's works form in themselves an ornithological museum; pictorial, we grant, but of such a character as to obviate the necessity of a collection of mounted specimens, obtained at no trifling cost, and preserved, even where room can be afforded for them, not without the greatest trouble."—*The Times.*

Gould's books on humming-birds, as well as the collection he formed of the birds themselves, which is now in the Natural History Museum, are known everywhere. The only pity is that his works are so inordinately expensive.

Besides such a magnificently standard work as this, embracing the birds of all countries, there are, passing eastwards, that ever delightful book, Captain Lloyd's "**Field Sports of the North of Europe.**"

"The passion for the chase is strong in Mr. Lloyd's constitution," writes a critic in *Blackwood's Magazine.* "It seems for years to have been his ruling passion, and to have made him a perfect model of perpetual motion. . . . We admire Mr. Lloyd. He is a fine specimen of an English gentleman; bold, free, active, intelligent, observant, good-humoured, and generous—no would-be wit, no paltry painter of the picturesque—above all, no pedant and philosopher. Mr. Lloyd's mind was wholly engrossed by his own wild and adventurous Scandinavian life; and when it was flown he then began to lead it over again in imagination."

His "**Game Birds and Wildfowl of Sweden and Norway,**" with an account of the seals and salt-water fishes (1867), is a valuable book, and should be possessed and its delightful plates studied by all interested in the summer homes of our various wildfowl.

To "An Old Bushman" (Wheelwright) we are indebted for an enticing picture, "**A Spring and Summer in Lapland,**" of collecting skins in the Lapland forests and witnessing the arctic winter vanish at the touch of spring.

That enlightened ecclesiastic, the Rev. Erich Pontoppidan, in "**The Natural History of Norway,**" has given a particular and accurate account of the temperature of the air, the different soils, waters, vegetables, metals, minerals, stones, beasts, birds, and fishes, together with the dispositions, customs, and manners of living of the inhabitants, interspersed with physiological notes from eminent writers, and transactions of academies, with map of Norway and 28 plates. He adds some information on fowling in Norway, with which I have occasionally made free.

Henry Seebohm's "**Siberia in Asia**" is full of curious facts regarding the migrations and nesting of English birds. Mr. Ernest Shelley, again, has written a comprehensive handbook on "**The Birds of Egypt,**" and a host of monographers, whom we have not space to detail at the length which their learning and research demands, have epitomized or amplified the feathered creatures of central and southern Europe.

The Indian sportsman keeps his "Jerdon" at hand, and cannot go far wrong while he has by him "**The Birds of India,**" in three volumes. There is also Le Messurier's "**Game, Shore, and Water Birds of India,**" though it is now very scarce; and Burton's "**Falconry in the Valley of the Indus,**" with four fine plates after Wolf and McMullin (1852); the "**Catalogues of the Birds**" in the Museum of the Hon. East India Company, by T. Horsfield, F.R.S., and F. Moore (1856); and others of various merits. The name of Mr. R. Bowdler Sharpe will always be held in high repute by Indian ornithologists. He has done much in classification or monographing, and there is at the present time perhaps no one more fitted, if he were willing, to prepare that urgently needed work, a clear, comprehensive, but concise, book on the birds of the Indian and Pacific Oceans.

In America there are good bird professors on every hand, besides sporting writers who compete with any in the mother country. The following are all useful books which may be consulted with advantage.

"**Game Birds and Water Fowl of the United States,**" 20 fine coloured plates, equal to drawings, each measuring twenty-two by twenty-eight inches, mounted on cardboard. List of plates: the American snipe, the green-winged teal, the woodcock, the mallard duck, the American quail, the black duck, the ruffed grouse, the blue-bill duck, the prairie chicken, the red-head duck, the Canada grouse, the wood duck, the Californian valley quail, the buffle-headed duck, the upland plover, the golden-eye duck or whistler, the Californian mountain quail, the widgeon, the canvas-back duck, and the brant; one volume, atlas folio (1878).

Wilson's "**American Ornithology,**" enlarged by Jardine, over 100 beautifully coloured plates of the birds of America, three volumes (1876).

"**Fauna Boreali Americana,**" the Zoology of the northern part of British America; the volume comprising the birds is by Swainson, illustrated by 52 coloured plates and wood engravings, royal 4to (1831).

Coue's "**Birds of the North-West,**" a handbook of the ornithology of the regions drained by the Missouri river and its tributaries (Washington, 1874).

"**The Birds of Jamaica,**" by P. H. Gosse.

Lewis's "**American Sportsman,**" containing hints to sportsmen, notes on shooting, and the habits of the game birds and wildfowl of America; and Long's "**American Wildfowl Shooting,**" containing full and accurate descriptions of the haunts, habits, and methods of shooting wildfowl, particularly those of the Western States of America; instructions concerning guns, blinds, boats, and decoys; the training of water retrievers, etc.; the true history of choke-bores, the theory of their action on the charge, construction, loading, etc., with a correct method of testing the shooting powers of shot-guns.

English game preserving has of late become a fine art. There was a time, and painfully remote it seems at present, when the only necessaries for a day's shooting, provided, of course, you kept off the king's manors and respected the abbot's fat bucks, were the implements of your craft with due skill. Now, alas, a day's shooting is a matter of solemn preliminaries, to which banker, solicitor, understrappers, and government licences are all accessories before the fact.

On game preserving as a means to a practical business-like result, Mayers, in his "**Park and Gamekeeper's Companion**," wrote in 1828; there is also Rawstorne's "**Art of Preserving Game**," and method of making plantation covers explained and illustrated, with 15 coloured drawings of shooting scenes, etc. (1837).

"**Practical Game Preserving**," containing directions for rearing and preserving both winged and ground game, and destroying vermin, with other information of value to the game preserver, by William Carnegie, is well known. "Mr. Carnegie gives a great variety of useful information as to game and game preserving, with many valuable suggestions. The instructions as to pheasant rearing are sound, and the chapters on poaching and poachers, both human and animal, are particularly to the point, and amusing withal."

Johnson's "**Gamekeeper's Directory**," with instructions for preservation of game, destruction of vermin, prevention of poaching, etc., is useful; and the author of the "**Amateur Poacher**" opens our eyes to many an artful device and ingenious wile.

Of books dealing with the art of approaching wildfowl there are: "**Hints on Shore Shooting,**" including a chapter on skinning and preserving birds, by J. E. Harting —an admirable little volume; "**The Dead Shot, or Sportsman's Complete Guide**;" a treatise on the use of the gun, dog breaking, pigeon shooting, etc., by Marksman, with plates; and Captain Lacy's "**Modern Shooter,**" containing practical instructions and directions for every kind of inland and coast work.

Of books on game laws, showing the keeper his relations to the poacher when his birds have come to maturity, there is Nelson's "**Game Laws of England,**" of hunting, hawking, fishing, and fowling, of forests, chases, parks, warrens, deer, dove-cotes, conies—a scarce and curious work; "**The Game Laws of England for Gamekeepers,**" by Hugh Neville, M.A., of the Inner Temple, barrister-at-law; and some few others. We may, however, safely say on this subject, that the epitome of English laws we have given in the following chapter possess the advantage of being unquestionably the most recent of any; and the summary of foreign game regulations in a final chapter has never, so far as we know, been attempted before.

If any one has a fancy for hawking, he may safely turn to the pictorial pages of the "**Falconer's Favourites,**" by W. Brodrick, a series of life-size, well-coloured portraits of all the British species of falcons used in falconry; or, "**Falconry in the British Isles,**" by Salvin and Brodrick, the second edition, with new plates and additions.

Finally, to bring our hasty and imperfect incursion into the realm of this literature to an end, the farmer who would know what English birds really eat all the year round should consult Napier on "**The Food, Use, and Beauty of English Birds**;" and the taxidermist, Rowland Ward's "**Sportman's Handbook.**" Other excellent manuals on this latter subject are Montagne Brown's "**Practical Taxidermy**;" Davies' "**Practical Naturalist's Guide,**" containing instructions for collecting, preparing, and preserving specimens of all departments of zoology, engravings, (Edinburgh, 1858); or, Kingsley's "**Naturalist's Assistant,**" a handbook for the collector and student, with a bibliography of 1500 works necessary for the systematic zoologist, illustrated, 8vo, cloth, Boston, 1882. One, N. Wood, has also prepared an "**Ornithologist's Text Book,**" a review of ornithological works, but it is long since out of date.

If the amateur bird stuffer, or the professional for that matter, would see and appreciate the highest perfection of this beautiful art, let him study the exquisitely arranged cases of the South Kensington Museum; or that splendid private enterprise, the Booth collection, in the Dyke Road Museum, Brighton.

Next to the endless pleasures of the open country and the studying of Nature as Gilbert White did, the companionship of wise and pleasant books is the naturalist's chiefest pleasure. Every one's taste or fancy will suggest certain books to him as more fascinating than others; but there is happily no lack of material in any direction, and, with a well and judiciously stocked library, he may still be cheerful when weather or unkind circumstances keep him from the active pursuit of his fascinating and ever soothing hobby.

## CHAPTER XI.

# GROUSE MOORS AND DEER FORESTS.

BY J. W. BRODIE-INNES.

SPORT in Scotland, according to its modern acceptation, presents many features peculiar to itself, and hardly to be found elsewhere; along with special fascinations, it has special difficulties and obstacles, which the English or American millionaire, who draws health and enjoyment from the heather hills, very imperfectly comprehends. In England, as in most other countries, sport has been a gradual development, whose direction has been determined partly by the nature of the quarry, and the facilities for breeding increased or lessened by the progress of agriculture in different districts, and partly by the invention and improvement of arms of precision, partly also by the gradual growth of the game laws; but, in the main, English sport to-day is the natural product and outcome of English sport centuries ago. In Scotland it is far otherwise. Within living memory the idea of the Highlands as a playground for the wealthy was unknown, and St. John's "Wild Sports of the Highlands" seems almost as archaic as Dame Juliana Berners. Within the memory of old men, such an event as a stranger coming to slay the grouse on the great barren hill-sides was very infrequent; no man bought or sold the game; the lairds and their friends shot for themselves and for presents. In the majority of cases the boundaries of properties were hardly known or heeded. If Seafield shot one hill, and Cluny shot another, no one knew or cared precisely where the line lay between them. Poaching there was, but it was for food or for sport, not for the filthy lucre of the city poulterer, and did but little harm to any one; neither did the sport of the lairds interfere in any way with the peasantry. The lot of the Highland peasant in those days was rough and primitive. Sheltered nooks in the hill-sides, where a turn of the hill protected a patch of decent soil, grew corn and potatoes enough to feed a family sparsely; a few hardly black-faced sheep supplied wool which the peasants themselves spun, wove, and dyed for their homely clothing; prices of grain and of mutton were good, if they had any to sell. No one dreamt of artificially keeping up a large head of game; and if damage were done to crops, it was more than compensated for by presents of game given liberally by laird or chief.

The opening up of the Highlands by railways and coach-roads, and the influx of tourists drawn thither by the fascination of Scott's novels, changed all the conditions of life as suddenly as the shift of a pantomime scene. For the peasants themselves, their lot had been grower harder, their struggle for existence more severe from many causes. Since they ceased to kill each other in constant clan feuds, and learned to live more healthy and sanitary lives, they rapidly increased beyond the capacity of the land to support them in anything like comfort; moreover, the natural indolence of the Celtic temperament led them to depend largely on the cultivation of the potatoe, and when the potatoe crop failed the congested district was plunged in misery and starvation. To these poor people the opening up of the Highlands brought the sharp contrast of comfort and luxury in city life, and the ready means of going thither, while the repeal of the corn laws, largely depressing the prices of produce, also had its necessary effect on a populace who were all vendors, and hardly, if at all, purchasers of articles of food. The concurrence of these and various other cognate causes began the depopulation of the Highlands long before the era of great sheep farms, grouse moors, or deer forests. Another resultant from the same great change was the discovery that the more delicate breeds of Cheviot sheep might with care thrive on the Scotch hills, and could be brought to perfection much earlier, and were therefore more valuable to the breeder than the hardy stock of former days. Then, by a natural sequence, came the large sheep farms in place of the deserted crofter townships. To assert, as is often done now, that the glens were cleared of men to make room for sheep, is to display the sheerest ignorance or wilful perversion of fact regarding the economic conditions. In a few instances this might have occurred, and in some cases no doubt tales of great hardship might be told; but in the vast majority of instances the people went voluntarily, or, if removed, it was to save them from a life of wretched dependence on charity, in a land which could no longer support them, even though they had it for nothing. But with the large sheep farmers came many wealthy Southrons eager to enjoy the sport of which they had heard so much; and as grouse and sheep lived amicably together, so the sheep farmer and the shooting tenant became corelatives, and the fascinations of grouse shooting grew into a fashion, and then into a craze, with startling suddenness; and thus the moors were parcelled out, and boundaries defined with mathematical exactness, and hosts of keepers and watchers employed to protect the dearly bought luxury. But it could not be expected that so sudden a revolution as this should all at once commend itself to the people, especially to a people so wedded to old tradition and old methods as the Scotch. Those who remained and had not joined the exodus to the towns, looked on the shooting tenants and the sheep farmers with a jaundiced eye; the thing was new, therefore abominable. The cry went up that the people were turned out for grouse and sheep. A few doctrinaires took it up, a few politicians for their own ends fostered it, and platform spouters, knowing no more of the Highlands than the interior of Africa, vapoured about it, till even some sensible people began to think there was some solid grievance; and thus sport in the Highlands grew up under the powerful stimulants of wealth and fashion on the one hand, and subject to the powerful opposition of political and social faction on the other.

No wonder the development was rapid, and the method of pursuing the *Tetras Scoticus* of Linnæus, or common red grouse, passed through numberless modifications in the course of a sportsman's memory. Few birds afford more delightful and exhilarating sport, followed as one used to follow them years ago, with the stout untiring English setters, over the purple moorlands, with many a knee-deep plunge in the soft boggy ground bordering the springs, where the grouse love to congregate, watching the clever systematic working of the dogs, and the point steady as a rock, when with a whirr and a rush a fine young cock rises perpendicularly some ten or twelve yards, then turns sharp for a horizontal flight, but at that instant, as he poises on the turn, the sharp challenge of the gun rings out, and a dishevelled mass of feathers lies on the heather. Such sport as this in the eye of the old sportsman cannot be excelled; but "*autres temps, autres mœurs*," the expenses of grouse-shooting have largely increased, the city poulterer gives a ready market for the quarry, and the temptation to make large bags, and so partially defray the expenses, becomes every year greater, though such an idea would have revolted the souls of the simple-minded lairds and chiefs of olden times, and is still looked on with great dissatisfaction by numbers of the peasantry. The invention and improvement of breech-loaders has tended to the same result, and conduced to the modern style of walking in line at short distances apart, with gillies following and carrying extra guns; till, in many parts, shooting over dogs is regarded as an antiquated amusement, fit only for old fogies. Whether arising from the frequent disturbance caused by this mode of pursuit, or from the larger head of game maintained on the moors, or from what reason it is hard to say, but certain it is that in an ordinarily fine season, when the birds are fairly early, by the beginning of September they grow as wild as hawks, and form themselves into large packs, either rising far out of gunshot, or sometimes to be seen running some five hundred yards away, ready to rise at the slightest step towards them, and fly a mile or more, only to pursue the same tactics again, should the sportsmen be ill advised enough to follow them. But the sport which has cost so much cannot be abandoned as hopeless after a fortnight or three weeks, and the old-fashioned plan of sending a steady old dog to head off the pack, till his master got near enough for a shot, would be far too tame and slow for modern ideas, and by no means productive of the big bags so much desired, and thus almost of necessity has come the practice of "driving"—little turf-built shelters concealing the sportsmen, who thus lie in ambush, while an army of beaters, marching across the heather, drive the grouse in flocks over their heads. A steady hand, a cool head, and a quick eye are all pre-eminently necessary for this mode of shooting, which, distasteful as it is to many of the old school, is by no means the cockney sport it is sometimes stigmatized as being. There is no catching the bird as he poises on his turn from the perpendicular to the horizontal flight; straight overhead, with a rush like an express train, goes the flight, the strong old cocks leading, and these it is the sportsman's aim to pick off, for it is well known that shooting down the old cocks is the best possible means of insuring a good stock on the moors in the following years. And thus the grouse drive has its own advantages and its own fascinations for the sportsman, though the comfortable shelters, the chairs, the luncheon, often attended by the ladies of the house-party, and served by elaborate flunkies, are apt to waken the disdain of old men accustomed to tramp for long hours behind a staunch dog, with nothing but a sandwich and a drop of whisky at the midday halt by the spring.

We have noted the dissatisfaction often evinced by the peasantry at the progress of Scotch sport, and the reasons for it, so far as those reasons amount to more than vague and formless discontent. There is no doubt that sheep farms and grouse moors, in some parts of the Highlands, occupy lands where in old time the crofter township drew scant subsistence from unwilling soil; there is no doubt that game of all kinds, under the fostering care of wealthy sportsmen, has enormously increased, and that the crops bordering on the great moorlands have suffered in consequence, while most of the game, which in old time found its way to the crofter's cottage, now goes to the city poulterer. Still, it must not be forgotten that, under the changed conditions of life, a crofter township, living as their fathers were content to live, in hardship and poverty, dependent merely on their own exertions to produce food, clothing, and shelter, is simply an Utopian dream. If the game laws were to be repealed to-morrow, the wild birds and beasts destroyed from off the great game-haunted hills, till the grouse became as extinct as the dodo, and the shooting tenants driven for their sport to Norway or Sweden, or some country wise enough in its generation to welcome them, can any sane man suppose that the glens would be forthwith peopled, as Lochiel well says, with "a happy and contented crofting peasantry, who would immediately show their satisfaction at the prospect presented to them of pastoral felicity and domestic comfort, by rushing into the arms of the first recruiting sergeant they might chance to meet?" On a moment's thought it must be obvious that, whatever platform-spouters may say, the existence of small crofting peasants depends on high prices of the produce they grow, combined with a simplicity of life and love of home, rendering them content with poverty and hardship, so only that they might stay in the land of their forefathers. These conditions are gone, never to return; only the love of home, deeply ingrained as it is in the Celtic nature, remains in a modified degree, and even that is now more often used as a lever to help an agitation than as a valid and living principle of action.

On the other side of the picture, the benefits resulting to the native population from the changed circumstances are obvious, however much Radical agitators may strive to prove the contrary. Let any one who knew the Highlands thirty or forty years ago pass through the country now. Who made those excellent roads, who built those trim shooting-boxes, and set up those miles on miles of fencing? Who but the peasants of the country, paid by the gold of the much-abused sporting tenant! Again, who watches and protects the game, traps the weasels and the stoats, the wild cats and the foxes? but the local peasantry, now finding congenial occupation as gillies and keepers. There are kirks for them to worship in, schools equal to any in Europe for the training of their children. Whence come the rates that provide these things? Once again, from the sporting tenant. Scotland, under free trade, and with all the competition of the world against her barren soil and her ungenial climate, can no longer support her sons by tillage; but Scotland as a play-ground, as a land of sport, as a producer of game, offers chances to her people, if they have but the sense to take them, such as few countries can vie with. Even now the government of Sweden are learning the lesson, and taking steps for the afforesting of vast tracts of their country, with a view to attracting some of the golden shower annually poured forth over the playgrounds of Europe; and they have consulted with the best experts in Scotland on the conditions most likely to ensure success. A movement like this forms a refutation of especial value to the sentimental vapourings of Professor Blackie and other political theorists, who would fain go back for a century, in respect of one feature in rural life, while retaining the habits, the requirements, the responsibilities, the moral and intellectual training which a century of progress has produced.

But further changes and developments have taken place

within the last few years. Improved communication with our colonies, and improved modes of conveying both live and dead meat and wool to the English markets, have greatly reduced the profits on sheep-farming; while the introduction, already alluded to, of the more delicate breed of Cheviot sheep by the South-country graziers, to replace the old hardy black-faced stock, has considerably increased the cost of production in many parts of the Highlands. Take, for instance, what are called the wedder farms, that is to say, land too high and rugged for breeding ewes, and there is a vast quantity of such land in many parts of the country, especially in the wild districts of Rosshire and Sutherland, or on the Grampian range. The stock on these farms consists of wedder lambs, put on the ground in August, and sold when three and a half years old. Some ten or twelve years ago the number of trains on the Skye railway, laden with sheep going in the early winter to the low country, was utterly astounding to a stranger; but the expenses connected with rearing and wintering this wedder stock have so increased of late, while the price of wool has fallen, that wedder farms are no longer profitable, and the South-country graziers have for the most part left Rosshire and Inverness, and even in the comparatively mild districts of Sutherland but few are now left. The results are serious in many ways. There are but three elements of value in Highland property, strictly so-called, viz. sheep-farm rents, sporting rents, and crofter rents. If economic conditions destroy the first, as has already happened in many of the districts, it is clear that, unless the sporting rents can be maintained, the whole burden of local rates and taxes must be borne by the crofters, and from what source the money for these purposes is to be obtained by men without capital, without experience or special skill, where those who command all these requisites have failed, it is hard to see. Even though the land were given rent free to the crofters, the rates and taxes necessary to maintain roads, police, schools, minister's stipend, etc., would far exceed any profit they could possibly hope to make. It is easy to say these expenses ought to be met by the landlords—*ex nihilo nihil fit;* and if all the value of Highland property be taken from it, whence is the landlord to satisfy the claims of the rate-collector? Happily, there is an alternative—where sheep cannot live, and where no blade of corn could be induced to grow is the favourite haunt and home of the great red deer, the noblest quarry that ever taxed the skill and endurance of a sportsman in the British islands. Probably there is no possible means whereby a wealthy man can secure a more abundant return, in health and enjoyment, for the money spent on an autumn holiday, than by renting a deer forest; in no other way can we account for the enormous sums spent annually on this sport, apart altogether from the sporting rent, which, as we have said, goes far to relieve the peasantry from the burden of rates and taxes. When we find that, in eighteen years, Mr. Fowler, of Bræmore, has spent £105,000, Lord Tweedmouth £50,000, and Sir John Ramsden £180,000, to take only three typical cases, and consider the classes of people among whom this money is spent and who benefit thereby, including masons, joiners, plasterers, plumbers, and slaters, with labourers for each trade, wire fencers, road-makers, blacksmiths, carriers, besides local shopkeepers, gillies, deer watchers, trappers, etc.; it must be evident that, so far from depopulating the Highlands, the creation of deer forests in fitting districts enables them to support a far larger population than under present conditions would otherwise be possible. The condition of the country under deer is widely different from that of moors under sheep and grouse. The latter, as we have seen, live amicably together. No special care is needed to avoid disturbing the grouse. Even in the breeding season the shepherds come and go, and the hen grouse will sit placidly on her nest and never heed them, when once she realizes that no harm is meant, and every enemy that can hurt the game is ruthlessly destroyed, whether it be weasel or polecat, wild cat or fox. On the deer forest, on the other hand, everything must be given up to the utmost quiet, so that the shyest and wariest of all wild animals may pasture in peace and undisturbed till the art of man meets the instinct of the animal in the attempt of the stalker to circumvent the noble stag on his own ground. Often and often has the sudden crow and whirring flight of an old cock grouse, startled by the tread of the deer-stalker, given a note of warning to the stag, maybe a mile or more distant, and spoiled a whole day's patient labour; often has some harmless tourist in search of ferns scattered a whole herd whereof probably he never saw or suspected a horn. Hence it is that owners of forests try their utmost to keep down the grouse, and to this end encourage the ground vermin, and forests generally swarm with the wild cat and the fox; and hence also trespassers are as sternly warned off as though the great bare hill-sides were a lady's pleasaunce. Very hard seems this latter restriction, so impossible is it for the ordinary non-sporting Sassenach to see where the harm comes in. Not a vestige of a deer can he see, nor does he for a moment understand or believe that his mere presence can make the slightest difference to any living creature on the distant hill face, whose purple heather looks to him merely an uninterrupted stretch of purple velvet, glowing in the sun and sending up its rich honey scents to the myriads of bees. But Donald from his shieling, watching through his telescope, sees peering above a heathery knoll a mighty pair of spreading antlers of ten points terminating in the orthodox three-pointed cup. Crouched down in that sheltered nook, with his harem of hinds keeping guard around him, lies the magnificent "royal" that shall be the prize of some wary stalker before the season is over; already his horns, clear of moss, are taking the brown hue like the peat bog. Donald measures them with his eye, and calculates how soon that noble head may be expected to hang in the hall of the shooting lodge, the finest trophy there; but as he watches, a feeding hind lifts her head with a quick gesture. What is the matter? Only the crackle of a dry scrap of heather under the feet of a spectacled professor with tin collecting-box, hunting for some obscure lichen—only this; but the hind's quick challenge spreads to her pasturing sisters, a dozen heads are lifted, dainty little hoofs stamp the ground, in an instant the mighty "royal" himself is on his legs. With a defiant toss of the great antlers, and a sniff at the breeze tainted by the presence of the poor meek professor, the whole group are off and away, rousing in their rapid flight other family groups, till to the keen eye of the old gillie the whole hill-side seems in motion; and, closing his telescope with a sigh, he recognizes the fact that, for a week at least, no sport can be had on that particular hill. It may be said, if the presence of a human being produces results so disastrous, how can the stalkers themselves go through the forest without spoiling their own sport? The answer is simple. At the commencement of a day on the forest the gillies and watchers have swept with their telescopes every nook and cranny of the hill the sportsmen are to try; every horn in sight is known and marked; the likely places for deer to lie *perdu* are noted; the direction of the wind and the turns and eddies with which it sweeps and swirls through the corries are carefully considered, and a line is chosen whereby, without alarming a single hind, the stag selected may be approached. Cautiously the little party creep from shelter to shelter, ever with an eye on the distant game; should a hind lift her head, the word is "*drop*," wherever you are, behind a rock, into a burn, it matters not; there you must lie till the alarm is past and the herd feeding quietly again. Often with wide circuits to avoid some obstinate cross current of wind that would bear the tale of your presence to the wary quarry, till at last, maybe after hours of patient clambering, creeping, lying hid, in short pitting your wits against the instinct and cunning of the keenest animal that lives, you are within rifle-shot; and now, beware lest the nervous agitation of the moment cause you to lose your head, or cause your hand to quiver and so you spoil that grand haunch, or worse still, commit the one unpardonable sin on a forest of wounding the deer. Better had you missed it altogether, better often that you should take the nearest train and boat to the wild west than appear before your host with such a confession.

But the average Englishman does not understand the conditions of sport on the deer forest. To him it seems grievous that the botanizing professor should be disturbed in the pursuit of his hobby; that the artist should be debarred from the lovely gleus; that the home-going peasant should be shut out from a short cut home across his native hills. And it seems grievous precisely in proportion to his inability to understand the damage these would do; and thus a door is opened to the political and other malcontents who would destroy the forests of Scotland, to do so by a side wind and indirect attack. The Access to Mountains Bill was a case in point. The amount of sympathy and support this proposal won was directly due to the impossibility of ordinary folk guaging the disingenuous nature of its proposals, and the specious fair-seeming with which they were brought forward.

J. W. Brodie Innes.

## The Laws of Covert and Fen.

Game and wildfowl laws, it may be fairly noted, are important in two respects. There is, firstly, their effect from a national point of view, their bearing on the abundance or scarcity of fur and feather itself. There are, secondly, considerations, dear to theoretical politicians, as whether the ground of a necessity devoted to them is wisely so devoted, whether they are good for the morality of the country-side, or whether they stir up hatred, and malice, and so on, in a circle of wide questions about which men have not agreed from the time when Nimrod cut off the ears of early Persian poachers down to to-day, which sees such matters provoking close divisions "in Parliament."

These questions are a science in themselves—a philosophy that is not to be summarized and dismissed in a few words. If, as is more than likely, the next fifty years sees England slowly revert to her condition in Saxon times and become an essentially pastoral country—a land of gardens, and orchards, and meadows, for all of them, statistics tell us, are slowly spreading while arable land is dwindling—then there is no reason why game of some kinds should not increase and become even more valuable than at present. This points the importance of the great issues which come under consideration when a populous nation with an inborn love of fresh air and field sports legislates for the delicate and susceptible fauna with which Nature has endowed its territory. There might well be a chair of ornithology amongst us, some learned and yet practical professorship where wisdom and observation on field matters might accumulate; but failing this a better general popular knowledge of game and wild birds is very highly desirable, not only in the curiculums of Cam and Isis, but even in village shrines of learning, and also the grimy benches of city schools where youthful devotees lisp their first homage to that triple-headed diety, the three R's.

The humbler sportsman's points of touch with legislation effecting him are few and simple. He needs his ten shilling licence "to carry and use a gun," obtained easily in England and Scotland, and with a little more formality in Ireland. This equips him legally, and between the 1st of August and the 1st of April he may shoot to his heart's content on the seas and estuaries; between high-water mark and low-water mark of mean tides (with some few and objectionable exceptions) on every beach all round the kingdom; as also on certain waste lands and warrens. The measure that limits his shooting season to the period of the year when birds are not breeding is short and concise. The following are its chief sections:—

THE WILD BIRDS' PROTECTION ACT.

§ 3. Any person who between the 1st of March and the 1st of August in any year after the passing of this Act shall knowingly and wilfully shoot or attempt to shoot, or shall use any boat for the purpose of shooting or causing to be shot, any wild bird, or shall use any lime, trap, snare, net, or other instrument for the purpose of taking any wild bird, or shall expose or offer for sale, or shall have in his control or possession after the 15th day of March, any wild bird recently killed or taken, shall, on conviction of any such offence before any two justices of the peace in England and Wales or Ireland, or before the sheriff in Scotland, in the case of any wild bird which is included in the schedule hereunto annexed, forfeit and pay for every such bird in respect of which an offence has been committed a sum not exceeding one pound, and, in the case of any other wild bird, shall for a first offence be reprimanded and discharged on payment of costs, and for every subsequent offence forfeit and pay for every such wild bird in respect of which an offence is committed a sum of money not exceeding five shillings, in addition to the costs, unless such person shall prove that the said wild bird was either killed or taken or bought or received during the period in which such wild bird could be legally killed or taken, or from some person residing out of the United Kingdom. This section shall not apply to the owner or occupier of any land, or to any person authorized by the owner or occupier of any land, killing or taking any wild bird on such land not included in the schedule hereto annexed.

§ 6. All offences mentioned in this Act which shall be committed within the jurisdiction of the Admiralty shall be deemed to be offences of the same nature and liable to the same punishments as if they had been committed upon any land in the United Kingdom, and may be dealt with, inquired of, tried, and determined in any country or place in the United Kingdom in which the offender shall be apprehended, or be in custody, or be summoned, in the same manner in all respects as if such offences had been actually committed in that country or place; and in any information or conviction for any such offence the offence may be averred to have been committed "on the high seas." And in Scotland any offence committed against this Act on the sea coast or at sea beyond the ordinary jurisdiction of any sheriff, justice, or justices of the peace, shall be held to have been committed in any county abutting on such sea coast or adjoining such sea, and may be tried and punished accordingly.

§ 9. The operation of this Act shall not extend to the Island of Saint Kilda, and it shall be lawful for one of her Majesty's Principal Secretaries of State as to Great Britain, and for the Lord Lieutenant as to Ireland, where it shall appear desirable, from time to time, upon the application of the justices in quarter sessions assembled in any county to exempt any such county or part or parts thereof, as to all or any wild birds, from the operation of this Act; and every such order shall be published and may be proved in the manner provided in the preceding section.

SCHEDULE.

| | | |
|---|---|---|
| American quail. | Kittiwake. | Sealark. |
| Auk. | Lapwing. | Seamew. |
| Avocet. | Loon. | Sea parrot. |
| Bee-eater. | Mallard. | Sea swallow. |
| Bittern. | Marrot. | Shearwater. |
| Bonxie. | Merganser. | Shelldrake. |
| Colin. | Murre. | Shoveller. |
| Cornish chough. | Night hawk. | Skua. |
| Coulterneb. | Night jar. | Smew. |

Cuckoo.
Curlew.
Diver.
Dotterel.
Dunbird.
Dunlin.
Eider duck.
Fern owl.
Fulmar.
Gannet.
Goatsucker.
Godwit.
Goldfinch.
Grebe.
Greenshank.
Guillemot.
Gull (except Black-backed gull).
Hoopoe.
Kingfisher.
Nightingale.
Oriole.
Owl.
Ox bird.
Oyster catcher.
Peewit.
Petrel.
Phalarope.
Plover.
Ploverspage.
Pochard.
Puffin.
Purre.
Razorbill.
Redshank.
Reeve or Ruff.
Roller.
Sanderling.
Sandpiper.
Scout.
Snipe.
Solan goose.
Spoonbill.
Stint.
Stone curlew.
Stonehatch.
Summer snipe.
Tarrock.
Teal.
Tern.
Thicknee.
Tystey.
Whaup.
Whimbrel.
Widgeon.
Wild duck.
Willock.
Woodcock.
Woodpecker.

By a rider to the above Act, added in 1881 (44 and 45 Vict. cap. 51), the "possessing" of protected birds in the close season is defined, and the lark is very properly added to the Schedule.

Whereas under section three of the Wild Birds' Protection Act, 1880, a person who within the period therein mentioned exposes or offers for sale, or has in his control or possession any wild bird recently killed or taken is liable to certain penalties therein mentioned, subject to the following exception, "unless such person shall prove that the said wild bird was either killed or taken, or bought or received during the period in which such wild bird could be legally killed or taken, or from some person residing out of the United Kingdom:"

And whereas doubts have arisen with respect to the construction of the above-recited enactment, and it is expedient to remove such doubts:

Be it therefore enacted by the Queen's most Excellent Majesty, by and with the advice and consent of the Lords Spiritual and Temporal, and Commons, in this present Parliament assembled, and by the authority of the same, as follows:

1. The above-recited exception in section three of the Wild Birds' Protection Act, 1880, shall be repealed, and in lieu thereof the following enactment shall have effect:

A person shall not be liable to be convicted under section three of the Wild Birds' Protection Act, 1880, of exposing or offering for sale, or having the control or possession of, any wild bird recently killed, if he satisfies the court before whom he is charged, either—

(1) That the killing of such wild bird, if in a place to which the said Act extends, was lawful at the time when and by the person by whom it was killed; or

(2) That the wild bird was killed in some place to which the said Act does not extend, and the fact that the wild bird was imported from some place to which the said Act does not extend shall, until the contrary be proved, be evidence that the bird was killed in some place to which the said Act does not extend.

2. The Schedule to the Wild Birds' Protection Act, 1880, shall be read and construed as if the word "Lark" had been inserted therein.

From these extracts the whole effect of the enactment can be judged, and it is, with this help, within the power of all who may be friendly disposed to the birds to assist in their protection.

Yet, simple as this "bill" is, we are constantly told it is full of errors of omission and commission. Some professional shooters declare its provisions may easily be evaded, with a bitterness, however, which suggests the measure to be singularly effective and useful in their part of the country! Then the flapper shooter says his young friends are strong on the wing and over-experienced on the threshold of August. He would like to get at them by the middle of July at the latest. The big gunner, in his punt off the Ipswich or Harwich flats, cannot understand why Parliament should cork up his four-bore just as the spring flight is on, and the ruffs and reeves are gambolling in an enticing manner on the marshes, and long strings of duck and wimbrel pass continually overhead. He argues without much logic that as they are going northward to "the foreigners" he ought to be allowed to take all he can reach before their departure. But far otherwise thinks the owner of decoy and snipe bogs. If all the birds *do* go northward in the spring he says it is chiefly owing to that hideous banging of seafowl ordnance going on "off Harwich." There was amorousness in the quack of the mallards and the bleating of the fen snipe even before the sallow buds were silky in early March, or the king-cups had put out a single new leaf to try the temperature of "the month that looks two ways." The better plan, according to this authority, would be to begin the close time with February—and especially as regards everything which puntsmen like to shoot. Mr. Morris, again, wants the gulls protected until September, and brings a strong case in his favour; but 'Arry, on the other hand, particularly desires them to be free food for powder and his borrowed gun when his August holiday turns him out to his own inclinations.

The Leadenhall poultrymen comfort themselves in knowing they may sell game from over seas when the sale of the same birds taken in Great Britain is forbidden. But this irks the tender ornithological compassion of men like the late Frank Buckland. On one occasion he wrote: "I have been consulted on a case which in the spring affects most seriously the supply of food to the public—namely, the importation and sale of capercailzie, blackcock, and ptarmigan in the English markets. Some game and poultry salesmen, of Liverpool, were summoned in March for selling these birds, and a nominal fine of a penny a head, and costs, was imposed. The penalties were inflicted under an Act of Parliament passed in the reign of William IV., cap. 32. This Act does not mention capercailzie; whether, however, the magistrates imposed fines respecting them I do not know. They, however, considered that the Act applied to birds imported from other countries. I have been asked to give my opinion upon this subject; I do so as a naturalist, but not as a lawyer. The London shops are at this moment full of capercailzie, blackcock, and ptarmigan. These are imported, *viâ* Hull, from Norway and Sweden, and enormous numbers of them are sent over during March and April from Bergen, Drontheim, and other ports on the west coast of Norway. Two questions now arise. First, is there any close time in Norway and Sweden for these birds; and, secondly, what is the actual condition of the birds as regards their state of (as we should say, if it were a question of salmon), spawning—*i.e.* nidification? I have made it my business to examine the internal anatomy of a female capercailzie, of a male and female blackcock, and of two female and one male ptarmigan, the birds themselves being much better witnesses as regards facts than anything recorded in books. I find in every case that the ovaries are exceedingly minute, and that, therefore, the birds are not yet near their breeding time. I find it recorded that the capercailzie go in packs during the winter, disperse in the spring, and nest about the beginning of May. The ptarmigan pair early in the spring, the eggs are begun to be laid in June. The blackcock nest in May. The above applies to the British Islands, the breeding in Norway and Sweden is probably later. As regards the law: In 1871 a most valuable report (C. 401) was presented to Parliament, giving the laws and regulations relative to the protection of game in eleven foreign countries. From this I learn that "for Norway and Sweden the legal season for killing deer, reindeer, capercailzie, hare, blackcock, hazel hen, and ptarmigan is from the 10th of August to the 15th of March." Provisions are also made as regards the young of useful birds or animals and the law of trespass. Such enormous quantities of game birds have been, and are now imported from Norway and Sweden to London, Liverpool, and other large towns, that I think we ought, as Englishmen, to inform the authorities of these countries of what is going on, in order that they may make inquiries into the effects that this spring slaughter may have upon their stock of game birds, and also into the manner by means of which such large numbers of these naturally shy birds are caught, especially as shot marks are rarely, if ever, found upon them."

Farmers hate many of those rapacious little songsters which compassionate ladies love, and naturalists and market-gardeners have never yet smoked a peace pipe together over the contents of a bullfinch or tomtit's stomach.

But this, at least, may be taken as certain, that these beneficent Acts are doing good on the whole; and this is, perhaps, as much as could be expected.

Regarding game, properly so called, there is as much contention and diversity of opinion. It is some satisfaction, however, to know this is no new thing, and that every civilized country on the face of the globe is protecting its more valuable bird life in maugre of political crotchet-mongers. These amiable gentry seize upon game as "the special luxury of the wealthy," and, consequently, a fit subject for their spleen. They would, if they could, sap the country gentleman's life of all its attractions, expatriate him, and distribute covert and woodland amongst their needy and dissolute following.

In 1831 the old territorial right in game was abolished, and with this was quenched all substantial grievances of these agitators, since everything worth shooting was henceforth at command of any one who cared to pay for the privilege in the most ordinary commercial fashion. But this measure, intended probably to diminish poaching by underselling the wood thief in his illegal booty, was ineffectual. It doubled the number of shooters, and swept away game from those freelands where it had hitherto existed in the security of the ample hedgerows of an early period, harried by few village firearms.

On the 1st of September, 1827, Lord Althorp, shooting by himself over unpreserved land in Warwickshire, where any one might shoot who pleased, killed twenty brace of birds to his own gun; a few days afterwards nineteen, one day fifteen, and two other days eleven brace. Such uncovenanted sport is, alas, utterly out of date; game has been accumulated into centres, and with its abundance in known localities and semi-domestication comes that chance of great booty to the poacher which Sir Robert Peel, alone of the statesmen of his time, foresaw when the thin end of the wedge was introduced by a well-meaning Liberalism in 1831. Then, again, the Rating Act of 1874, for the first time taxed sporting rights as such; and more recently Sir William Harcourt's Bill gives to the tenant the right of killing the hares and rabbits on his own farm, any agreement with his landlord to the contrary notwithstanding: that is to say, that no covenant on the occupier's part to reserve the ground game for the proprietor can any longer be enforced by law. The game in England, be it remembered, whether four-footed or winged, had always belonged to the tenant, and only ceased to be his when he transferred it by agreement to his landlord. This transfer he is now, as far as ground game is concerned, forbidden to make, and can no longer therefore do what he will with his own.

A clause in this latter enactment makes it necessary to set rabbit traps (by the tenant who does not own the game shooting) only in rabbit holes, or runs. A learned judge, I notice, has given it as his opinion that a "run," or burrow, ends with a line drawn perpendicularly from its upper outer edge.

Thus, as matters stand at present, hares and rabbits may be shot by the occupier of the land, while pheasants, partridges, and a few other species of legal "game," are the property of the person in whose hands the shooting is.

Living, much as I have done, in various counties, and seeing, as it has been my good fortune to do, various kinds of game manors, from the roughest to the most scientific, I do not think that there is a shadow of justification for the outcry against game preserving, except perhaps on the single count that it encourages poaching. To suppress it on this account would be as reasonable as to mollify the nocturnal burglar by smelting down your silver spoons, or to content yourself with one change of linen per annum out of sympathy with the wardrobes of the Great Unwashed! As it has been well said by a reviewer of these laws: "What really underlies the Radical outcry is, not compassion for the ill-used agriculturist, but jealousy of the territorial magnate. It is really the sporting right, and not the game, which all the hubbub is about. The farmer, when he grumbles at all, grumbles for want of the shooting. The righteous indignation of the Radical is really inspired, not by the sight of rabbits nibbling the wheat, but of a country gentleman with a gun, suggesting to his diseased imagination ideas of Front de Bœuf or William Rufus, and sending him back to the commercial room of his hotel to startle all who hear him with his pictures of rural tyranny, patrician insolence, downtrodden serfs, and all the other well-known abominations of 'landlordism.'"

The agricultural labourer of the day has no doubt substantial troubles. When he is told that there is a panacea for all these in the tender and effusive affection of Socialistic agitators, he is cozened, against his homely good sense, into supporting the side of big promises, and, perplexed by the confusion of the day's burning questions, hearkens to the teaching of Will o' the Wisps. To begin with, his stomach is empty, and this is a condition utterly at variance with the cheerful belief in those good times to-morrow which some counsel him to accept in place of dinner to-day. He wants well-paid work, that he may live as comfortably as he lived thirty years ago; not the miserable occasional job—a mockery of steady employment—too often marking the condition of the market in rural districts. The labour representatives, or rather the representatives of labour discontent, tell him that wealth is stagnant in the social spheres above, and if the land is to be fertilized, it must be by such a golden shower as would result from puncturing the money-bags of the wealthy. The attributes of opulence, aggressive everywhere in this fair and delightful land, seem to endorse the crude logic of these democrats. Even such follies as "game for every one," or "three acres and a cow," are not above the hungry wonder of Hodge, whose little ones, in a land overflowing with milk and honey, pine on "skimmed Simpson" as blue as ever disgusted a cockney, and unpaid-for bread from speculative village bakeries "as dry as the remainder biscuit after a voyage" which Jacques scorned—and far less wholesome! Nor is it politic to whittle those privileges which were but unquestioned rights of "rude forefathers of the hamlets." Those countrymen are but mortal after all, and are full of the passions of their kind, the egotism, and even the pride which Pope suggests marks a fool equally in fustian or broadcloth; they feel the tight grip which modern competition has put upon every scrap of land, more, perhaps, than some of their friends know; and it saps their belief in the kindly fellowship of those above them when the valleys blossom with forbidding notice boards, and coppice and common, where children played and winter firewood came from, glisten in the sunshine a maze of steel-barbed fencing.

This is the tune to which the agitator of more pay and less work for horny-handed sons of toil tunes his fitful lute.

He whispers in the ear of any one who will listen to him that while the squire basks by the glow of Wallsend fires, his good friend the cotter—let the winter be never so Siberian—must not roam the park and garner the fallen timber that lies rotting there. Why, even water has been misappropriated, explains the virtuously indignant champion of the rustic, pointing the finger of hatred at hamlets where the village pump has been run dry that great folk may make ponds on their front lawns for foreign wild-fowl! The shafts in his quiver are many, and he uses them adroitly; he points to the Irish labourers who swarm across the shallow seas and send down the value of the Englishman's labour even in his own fields. If Home Rule were granted, is one argument among many, these frugal gentry would keep within their own bogs and cotton-grass wastes, and that alone would be a strong advantage. All this hoodwinks the youngest among our hinds. The elders among them, as far as we have observed, are more circumspect in their opinions. They hear without enthusiasm tall talk of abolishing game laws and throwing open coverts to general pillage; the more thoughtful know that all the pheasants and partridges in the kingdom would not go twice round amongst our population, or flavour the rustic's evening pottage even for a week. Nor are they in favour of general confiscation who live by the judiciously placed capital of landowners, and the ceaseless need of all classes for corn and beef.

What those who stand by the land *do* need to keep them in the political way they should go, is, firstly, "better times," which no party, alas, can create; and, secondly, the countenance and goodwill of those whom chance has dressed in a little brief authority. A landowner indeed, we strongly feel, and the hind feels too, who lodges his horses better than his husbandmen, and loves his orchids better than those chubby children who are to serve and live by the side of his heir, no more deserves to legislate for his district than he deserves to be honoured in it.

The farmer, again, may dislike to see total strangers perambulating his beans and clover, but "cussin" partridges or the shooting tenant is poor agriculture at best, and what he wants is not what the Radicals prescribe.

Mr. Brodie Innes' admirable sentences, prefaced to this chapter, shows where the shoe pinches in the north, and the real position of the question there. For myself, I will not attempt to epitomise the fine subtilties of "1 and 2 Will. IV. cap. 32," or "24 and 25 Vict. cap. 96," or, indeed, any other chapter or heading whatever, though there is a goodly pile of tomes devoted to this literature at my elbow. Already there is a keen desire on foot amongst the sensible yeomanry of the midlands to amend the Ground Game Act, and give the much persecuted hares a close time. Winged game was never more plentiful or better appreciated than it was last season; and if the agriculturists can be got to see that the abolishment of game is but a selfish propaganda prettily bound—a plausible repetition of Metternich's formula, "Ote-toi de la, que je m'y mette," all will be well, and we shall drop no substantial possessions for very shadowy and more than mythical advantages.

# CHAPTER XII.

# *GAME LAWS ABROAD.**

## GERMANY.

As the law now stands, any person in Prussia owning not less than two hundred English acres of land together, and who procures annually a game certificate at a cost of three shillings (in Hanover it costs six shillings, and in Hesse nine shillings), has an unrestricted right to kill all game upon his own property, and the same right is

* *List of Reports* (*C.* 310, 0/6–1871).

| Country. | Residence. | Name. | Date. |
|---|---|---|---|
| DENMARK ... ... | Copenhagen ... | Mr. Strachey ... | November 8, 1870. |
| NETHERLANDS ... ... | The Hague ... | Vice-Admiral Harris | January 28, 1871. |
| PERSIA . ... | Tehran ... ... | Mr. Jenner ... ... | October 8, 1870. |
| PORTUGAL .. ... | Lisbon ... ... | Mr. Doria ... ... | December 28, 1870. |
| PRUSSIA ... ... ... | Berlin ... ... | Mr. Petre ... ... | February 6, 1871. |
| RUSSIA ... .. ... | St. Petersburg | Mr. Rumbold ... | February 2, 1871. |
| SPAIN .. .. ... | Madrid ... ... | Mr. Ffrench ... | December 12, 1870. |
| SWEDEN AND NORWAY | Stockholm ... | Mr. Jerningham ... | November 30, 1870. |
| SWITZERLAND ... ... | Berne ... ... | Mr. Bonar ... ... | September 24, 1870. |
| TURKEY ... ... .. | Therapia ... | Mr. Moore ... ... | August 30, 1870. |
| UNITED STATES ... | Washington ... | Sir E. Thornton ... | January 23, 1871. |

extended to all enclosed lands of whatever extent they may be. Unenclosed properties of less than two hundred acres do not entitle their owners to kill the game on their own lands; these revert, for all sporting purposes, to the commune in which they are situated, and form a common shooting district. The communal authorities are bound either to appoint a gamekeeper to shoot over the district, or to let the shooting, or to leave it in abeyance; in either of the two former cases the profits derived from it are divided between the owners of the lands which form the district. An exception to this rule is made in the case of properties of less than two hundred acres which are situated in the midst of, or are partially surrounded by a forest of more than two thousand acres in extent, which is in the possession of a single owner. In such cases the owner of the land, instead of annexing it, as he would be compelled to do under ordinary circumstances, to the communal shooting district, is bound to let the shooting to the proprietor of the surrounding forest. Should the latter decline to avail himself of this right, the landowner may kill the game himself; or, if they are unable to agree as to the terms of the lease, the landrath is called in to arbitrate. The right of shooting upon all lands owned by corporations, or by more than three joint proprietors, must either be delegated to a gamekeeper or leased to a tenant.

As regards compensation for damages caused by game, it will be seen by a reference to the 25th Section of the law that no legal claim whatever can be preferred in Prussia for indemnity for any loss or injury incurred under this head. Under the old laws of Prussia, at a time when the right of shooting was separated from the possession of the soil, a landowner whose crops were damaged by the excessive preservation of game, was entitled to compensation for the injury inflicted; but this is now no longer the case, although the law sanctions or enjoins certain indirect means of counteracting, or rather of mitigating, the evil.

By another important provision of this game law, the sale of home or foreign game is prohibited from fourteen days after the expiration of the season, during which game may be lawfully killed.

The provinces where game most abounds, excluding the newly annexed territories, are Prussia, Silesia, Brandenburg, and Saxony. Herr von Hagen estimates as follow the quantity of game annually killed in the provinces of Prussia, together with the number of pounds of meat which it produces and its money value:—

| | No. | | lbs. | | lbs. | | Silbergroschen |
|---|---|---|---|---|---|---|---|
| Red deer ... ... ... | 4,288 | at | 120 | = | 514,560 | at | 2½ |
| Fallow deer ... ... ... | 2,546 | „ | 50 | | 127,300 | „ | 2½ |
| Roe deer ... ... ... | 14,204 | „ | 25 | | 255,100 | „ | 4 |
| Wild boars ... ... ... | 2,358 | „ | 60 | | 141,480 | „ | 3 |
| Elks ... ... ... ... | 54 | „ | 250 | | 13,700 | „ | 1½ |
| Hares ... ... ... ... | 1,097,316 | „ | 5 | | 5,486,580 | „ | 3 |
| Partridges ... ... ... | 1,311,134 | „ | ¾ | | 983,351 | „ | 5 |
| Pheasants ... ... ... | 2,373 | „ | 2 | | 4,746 | „ | 10 |
| Black game ... ... ... | 1,340 | „ | 2 | | 2,680 | „ | 7½ |
| Hazel game ("Hazelwild") | 992 | „ | ¾ | | 744 | „ | 10 |
| Snipe ... ... ... | 13,132 | „ | ½ | | 6,566 | „ | 10 |
| Wild ducks ... ... ... | 16,454 | „ | 1½ | | 24,681 | „ | 3 |
| Rabbits ... ... ... | 8,308 | „ | 2 | | 16,616 | „ | 1 |
| Fieldfares ("Krammetsvögel") | | „ | | | | | |
| "schock" of three score | 4,824 | „ | 15 | | 72,360 | „ | 2 |
| Total ... ... ... ... ... | | | | | 7,750,464 | | |
| Of the value of ... ... ... | | | | | 840,752 thalers. | | |

To the money value is to be added—

| | | |
|---|---|---|
| 11,524 foxes, at 1 thaler the skin ... ... | 11,524 | thalers |
| 643 badgers, at 2 thalers, ditto ... ... | 1,286 | „ |
| Hides and skins of red deer, at 1⅓ thaler ... | 5,717 | „ |
| „ fallow deer, ⅔ thaler ... ... | 1,697 | „ |
| „ roe deer, ⅕ thaler ... ... | 2,841 | „ |
| „ elks, 3 thalers ... ... | 162 | „ |
| „ wild boars, ½ thaler ... ... | 1,179 | „ |
| Hare and rabbit skins, 3 groschen ... ... ... | 110,562 | „ |
| Total value ... ... | 975,720 | „ |

Equal to £146,358 sterling.

§ 1. The fence periods are:—

1. For the elk, from December 1 to August 11.

2. For male red and fallow deer, from March 1 to the end of June.

3. For female red and fallow deer and fawns, from February 1 to October 15.

4. For roebucks, from March 1 to the end of April.

5. For does, from December 15 to October 15.

6. For roe calves (fawns), the whole year.

7. For badgers, from December 1 to the end of September.

8. For capercailzie (cocks) ("Auerhähne"), blackcocks ("Birchhähne"), and cock pheasants, from June 1 to August 31.

9. For wild duck, from April 1 to June 30; the fence time may be abolished in particular districts by the provincial governments.

10. For bustards, snipe, wild swans, and all other fen birds and water fowl, with the exception of wild geese and herons, from May 1 to June 30.

11. For partridges, from December 1 to August 31.

12. For hen capercailzies, grey hens, and hen pheasants, hazel game ("Hazelwild" or "Gelinottes"), quails, and hares, from February 1 to August 31.

13. It is forbidden all the year round to snare partridges, hares, and roe deer.

All other descriptions of game, including cormorants, divers ("Taucher"), may be taken or killed the whole year round. The young of red, fallow, and roe deer are to be considered as fawns ("Kälbe") up to the last day of the month of December following their birth.

§ 2. With a view to the protection of agriculture and to the preservation of game, the provincial governments are authorized to fix otherwise each year by special order the period at which the fence season for the description of game specified in §§ 7, 11, and 12 is to commence and close; but so that the fence time shall not commence or close more than fourteen days before or after the time fixed by § 1.

§ 3. The legal rights which exist in particular districts in respect to killing game even during the fence time, as a protection against damages caused by game, are not affected by the present law.

§ 4. The present law does not apply to killing game in enclosed parks. But the sale of game killed in such parks during the fence time is prohibited in conformity with the provisions of § 7.

§ 5. The following are the fines incurred for killing or taking game during the fence periods, as also for trapping or snaring game:—

| | Thalers. | | £ | s. | d. |
|---|---|---|---|---|---|
| 1. For an elk ... ... ... ... | 50 | = | 7 | 10 | 0 |
| 2. For a red deer ... ... ... ... | 30 | | 4 | 10 | 0 |
| 3. For a fallow deer ... ... ... ... | 20 | | 3 | 0 | 0 |
| 4. For a roe deer ... ... ... ... | 10 | | 1 | 10 | 0 |
| 5. For a badger ... ... ... ... | 5 | | 0 | 15 | 0 |
| 6. For a capercailzie (cock or hen) ... | 10 | | 1 | 10 | 0 |
| 7. For a blackcock, or hen ... ... ... | 3 | | 0 | 9 | 0 |
| 8. For a hazel cock ("Hazelhahn") or hen | 3 | | 0 | 9 | 0 |
| 9. For a pheasant ... ... ... ... | 10 | | 1 | 10 | 0 |
| 10. For a swan ... ... ... ... | 10 | | 1 | 10 | 0 |
| 11. For a bustard ... ... ... ... | 3 | | 0 | 9 | 0 |
| 12. For a hare ... ... ... ... ... | 4 | | 0 | 12 | 0 |
| 13. For a partridge ... ... ... ... | 2 | | 0 | 6 | 0 |
| 14. For a snipe, wild duck, or any other species of water fowl included under the head of game ... ... ... | 2 | | 0 | 6 | 0 |

§ 6. It is forbidden to take up the eggs or brood of game birds, and the prohibition extends even to persons to whom the shooting belongs; the latter, however (in particular the owners of pheasant preserves), are authorized to take up the eggs which are laid in the open, in order to have them hatched.

It is equally forbidden to take away plovers' and seagulls' eggs after April 30.

§ 7. Any one hawking, or exposing, or offering for sale in shops, markets, or in any other way, game, whether entire or cut up, but not cooked, the taking or killing of which is prohibited at the time, fourteen days after the commencement of the fence time, or any one assisting in such sale, incurs, besides the confiscation of the game, a fine not exceeding 30 thalers, to be applied to the benefit of the poor-box of the commune in which the offence is committed.

### Russia.

Paragraph 535 of the regulations respecting shooting states in general terms that every landowner has the right to shoot on his own lands, and on lands rented from the crown ("Kazonnia Zemli"), subject to certain restrictions as to the time of year when such right may be exercised.

Par. 536 states that shootings on the lands of others is permissible only with the written authorization of the owner.

Pars. 537 and 538 prohibit the driving of beasts and birds out of lands belonging to others, as also damaging places to which birds resort on their flight ("ptitchi privali"), or removing and carrying away traps, snares, etc., used for catching birds and beasts. Birds' nests are not to be destroyed, nor the eggs carried off. The only exception applies to the nests of birds of prey.

From 1st of March to St. Peter's day (29th June, O.S.) it is strictly forbidden to shoot birds or beasts, both on private lands and crown lands, or to catch game in pits, nets, nooses, traps, or by the means of any other instrument whatsoever. In the governments of St. Petersburg, Novgorod, and Pskoff, the period of prohibition is extended to the 15th of July, O.S., and an exception is made in respect of blackcock and capercailzie, which may be lawfully shot in the spring during calling-time. With a view to prevent the wanton destruction of game in spring time, it is severely prohibited to carry into the towns and there to sell any kind of game from the 1st of March to the 1st of July, O.S. The town and district police, as well as the starosts or other village authorities, are bound to see this rule carried out. It may, however, be observed here that it would not appear to be successfully enforced.

Pars. 545 and 546 state that wild beasts and birds of prey

are excepted from the above prohibition, and that it is lawful to destroy them in any manner or at any time of year. Within the prohibited period above mentioned, only land-owners, however, and the gamekeepers in their employ, are allowed to destroy wild beasts on their private lands without first giving notice to the local police. All other persons intending to destroy wild beasts (such as bears, lynxes, wolves, foxes, etc.), within that period, must give preliminary notice of such intention.

Par. 763 treats of shooting licences. These are said to be required of all persons shooting in the neighbourhood of the capitals (St. Petersburg and Moscow), even during the lawful shooting season. They are issued by the department of the Ober-Jägermeister, and any person found shooting without one is liable to fines, and in case of non-payment of such fines, to the confiscation of his gun and dog or dogs. The evasion of this regulation out of the shooting season exposes the offender to the confiscation of all his shooting implements, in addition to the fines in money laid down in Article 1172 of the Statute of Punishments. All these fines and confiscations to go to the Jägermeister department aforesaid.

### Sweden and Norway.

On leasehold property the tenant is entitled to kill game, unless it be otherwise stipulated in the lease.

In the northern provinces much land is totally unsurveyed, and unapportioned to any one; there exists also there what is termed "Overlopp's land," or land in excess of what forms the proper area of a homestead, as determined by the official survey. This latter land falls by law to the crown, and is employed in augmenting such homesteads as are deficient, and in creating new ones.

On these two descriptions of land the right to kill game and wild beasts is enjoyed by every one; subject, of course, to certain restrictions.

The destruction of elk, however, is here again prohibited, without the royal sanction.

On disputed lands, which are still under litigation, neither of the contending parties is allowed to pursue game, although vermin and noxious animals may be killed.

All game wounded by the owner, or persons authorized by him, on his own property, may be pursued and taken upon a neighbouring property; and should the animal so hunted be a bear, wolf, lynx, or glutton, the above provision holds good, although it be not previously wounded.

Any person who has surrounded and marked down a bear in its den during winter shall enjoy the sole right of pursuing and killing it; and no one, not even the owner of the land, shall be entitled to disturb the beast, or to prevent the chase.

Should any person undertake to destroy any of the above-mentioned wild animals in any other manner on land not his own, but where they are known to exist, he must give notice of his intention to the landlord, who is entitled to participate in the hunt; but in no case can he prevent its taking place, the notifier appropriating the animal, and any one, no matter who, meeting with beasts of prey on any land whatever is entitled to kill and keep them.

The following are considered, according to Swedish law, to be vermin and beasts of prey: bears, wolves, lynxes, gluttons, foxes, martens, otters, seals, eagles, eagle owls, hawks, and falcons.

Certain rewards are to be paid from the public treasury for the destruction of beasts of prey: fifty dollars (nearly £3) for a bear; twenty-five dollars for a wolf or lynx; and ten dollars for a glutton; the destruction of the young of these animals receiving the same recompense.

The legal seasons for killing game are as follows:—

Elk may be killed from 10th of August to 1st of October.

Beaver from 10th of July to 1st of November.

Partridge and grouse from 1st of September to 1st of November.

Swans, wild ducks, eider-ducks, snipe, and woodcock, from 10th of July to 16th of March.

Deer, reindeer, capercailzie, hare, blackcock, hazel-hen, and ptarmigan, from 10th of August to 15th of March.

The above are the seasons as established by the latest law on the subject in 1869; but local regulations exist in the various magisterial communities throughout the country, which modify its provisions in a slight degree, and these lengthen or shorten the legal periods, according to the habits of the different kinds of game which frequent the several localities.

As may be readily conceived, these are widely different in a country extending over so many parallels of latitude as Sweden.

The eggs of feathered game are also protected by law; it being illegal to rob any nest, or to destroy the young of any of the above-mentioned animals, before the 10th of July, or of any other useful bird or beast, before the 10th of August.

There is one important difference between the British and Swedish game laws; for, in any enclosed hunting-ground or park, it is lawful, at all times of the year for those who enjoy the right from the owner, to kill any species of game or wild animal found therein, so that the above regulation as to season virtually only affects those persons who pursue game upon unenclosed land.

It is not permitted to offer game for sale during the prohibited periods of the year, unless legal proof can be given that it has been killed lawfully, or upon enclosed land.

Snares, with spears attached, and spring-guns, for the destruction of game, are illegal; and elk may not be hunted on skates, or taken in pitfalls.

Ordinary traps and snares for killing game and wild animals, cannot be used by a person on land not his own, without the consent of the owner; and they may in no case be set out between the 31st of May and 1st of October; and when laid out, they must be guarded from the approach of domestic animals. Notice of the intention to use such traps, etc., being read out in the parish church once a month until their removal.

Persons shooting or hunting any of the above-named animals at unlawful times are liable to a fine of from ten to two hundred dollars; and should the animal be an elk, or beaver, the fine is to be not under one hundred and fifty dollars.

The same penalties apply to persons offering game for sale at illegal times.

Any person pursuing ordinary game at illegal periods may be deprived of his game, and also of his guns, dogs, and sporting appliances, by any person discovering him, and the property may be retained until the case shall have been judicially investigated.

Should the offence thus committed be solely against the rights of private individuals, they alone shall be entitled to prosecute; but should any of the legal enactments for the benefit of the public have been transgressed, the offender shall be prosecuted by the public accuser, or by a member of the Board of Woods and Forests.

One-third of all fines inflicted under the game laws shall belong to the crown, two-thirds going to the informer.

Any one unable to pay the fines is liable to a proportionate period of imprisonment according to the Swedish Penal Code.

All forfeited game is to become the property of the informer.

The above laws are stringently enforced in most of the provinces where game exists in large quantities; and the clauses relating to the preservation of elk and beaver are but seldom infringed, the highest penalty which the law permits being in all cases exacted.

It will be observed that no game licence or gun-tax is payable.

It is no uncommon thing for large hunting-parties to come together for the destruction of bears, wolves, and lynxes infesting the districts of Vermland and Norrland, and which become a source of annoyance and even danger to the scanty population of those northern provinces.

Bears, indeed, are not frequently killed except on such occasions, and the reward offered by the Government is in these cases not given, the personal danger incurred being so much lessened.

Wolves, however, in severe winters, approach the large towns in search of food, and the sums paid for their capture are often considerable.

### Holland.

The right of shooting, coursing, fishing, or any other kind of sport, is attached in this country exclusively to the ownership of the land; the game is looked on as a natural production of the land, in the same way that the fish is regarded as a natural production of the water, the heather of the heath, or the tree of the forest. Where the land belongs to the private individual, the game which is on it is private property; and where the land is the property of the State, to the State belongs also the game upon it. And in like manner, as it is in the power of the landowner to let out his property to another to be cultivated, while he reserves to himself or concedes to a third party the right to cut the timber, so he is at liberty also to introduce clauses into the lease reserving the right of shooting the game himself or leasing the shooting to another.

The principle that game is the natural property of the owner of the soil is so thoroughly recognized that the legislature has practically decided that any alienation of the one from the other is an unnatural one; and, in cases where such alienation has occurred, has given to the proprietor the right of shooting, even though in direct opposition to a written covenant.

But although the right of a landed proprietor to the game on his estate is regarded as a natural and an inalienable one, it is in its nature imperfect. Though his ownership is in one sense absolute, the mode in which he may derive advantage from it is restricted. Numerous regulations exist, having for their object the preservation of the game, and with such he must comply. The shooting season is limited in length; shooting on Sunday or at night is strictly prohibited, as also during those times when snow is lying on the ground, or the land is flooded; the mode of capturing or killing game is strictly defined, nets, traps, and snares being in nearly all cases prohibited; the number of dogs to be employed in coursing in the same field, and the amount of slaughter to be committed in the same battue are also limited. It is likewise forbidden to transport or sell game during close time.

The State takes upon itself, to a great extent, the duty of preventing poaching. This it does with the aid of various regulations, such as preventing persons other than the owners shooting on land without written permission from the owners, or even being found in a field or wood with a gun; and, again, making it an offence to convey game from one place to another except by the public road or footpath, or without a written certificate showing the means by which the game was obtained. Even in the case of game introduced from abroad, it cannot be removed from the port of entry without a certificate of its foreign origin.

The right of shooting cannot be alienated from the ownership of the property.

Art. 11. The committee of the provincial states shall annually fix the time in each province for the opening and closing of the shooting season, as well as the days of the week when small or large game may be killed, and the commissary in the province shall give notice thereof at least eight days before the opening and the closing.

In like manner the committee of the provincial states

shall decide, according as the condition of the game or local circumstances require, whether shooting any particular game . . . shall be forbidden or limited, either over the entire province or in certain districts, as well as how many head of large game, male or female, may be killed, and how many hares may be shot or taken in one day by one person, or how many in a single battue; and furthermore, they shall appoint the time during which the decoy ducks shall be shut up.

Art. 12. No licence or other special authority required—

(*a*) To permit the owner or other rightfully empowered person to shoot in pleasure grounds, gardens or other grounds enclosed by walls, screens, fences or canals.

(*b*) To permit the shooting of destructive birds in gardens or orchards by the owner or other rightful person, or by his order.

Art. 18. Shooting is prohibited—

(*a*) On Sundays.

(*b*) Before sunrise and after sunset, with the exception of the pursuit of such game as is referred to in Article 15, letters *e*, *f*, *g*, and *h*, as also of duck shooting, which are permitted for half an hour before sunrise, and half an hour after sunset.

(*c*) In time of snow, with the exception of the battues referred to in Article 16, and of waterfowl shooting on the sea-shore, and on the banks of rivers, marshes, etc., and of the pursuit of such game as is referred to in Article 15, letters *g* and *h*.

Art. 22. It is forbidden to seek, pick up, sell, expose for sale, or transport the eggs of game.

Art. 27. Selling, exposing for sale, or transporting game in close time is prohibited, except for fourteen days after the closing.

Even in the shooting season it is forbidden to carry or transport game in fields, or away from public roads and footpaths, unless the person himself who is carrying the game, or some one accompanying him, holds a licence, or unless he holds a "permission-gratis" from the burgomaster of the parish where he resides, to be produced on the first demand of the properly qualified officer.

Art. 45 condemns to seizure and confiscation the guns and other implements (but not dogs) employed by any one shooting or pursuing game in close time or other forbidden times, or without a licence, or without permission from the owner, or pursuing game in an unlawful manner. It likewise orders the confiscation of the game unlawfully killed, exposed for sale, or removed. The offender has, however, with certain exceptions, the option of retaining the objects confiscated and paying their value. In case no seizure has been actually effected, the offender has still to pay their value as estimated by the magistrate with the aid of evidence.

It will be seen from the above that the game law of this country resembles in many respects those of the British Islands, and that where it differs it is generally in the sense of greater protection to the game and more numerous restrictions on the sportsman.

## AMERICA.

Before entering upon a consideration of the laws and regulations throughout the United States which relate to the protection of game and to trespass, it must be stated that no general or uniform law governing the whole country exists on either subject. Legislation on these and on kindred matters lie beyond the domain of the Federal Congress, and depends entirely on the legislature of each separate State. In order, therefore, to arrive at the required information, it has been necessary both to consult the several statute books of the thirty-seven States, and also to make inquiries as to the common law obtaining in different parts of the country. This having been done, it has been ascertained that whilst in every State there exist laws regarding trespass, it is only in twenty-nine of them that enactments have been passed for the preservation of game; although those few States which have not legislated upon the latter subject are among the least important and the least populated in the Union.

It must first be remarked, then, that in their titles the laws always profess to be, not for the protection of game as for the profit or enjoyment of the proprietors of land, but for its preservation as for its popular and general use. Notwithstanding, however, this evident interest of the different Acts that the legislation on this subject should be for the protection rather of public than of individual interests, there is not the slightest indication that the game on private lands is to be considered the property of the State, or of any other person than the landlord.

There is no law in any State of the Union requiring a game certificate or a licence for carrying a gun.

Everywhere it is forbidden to shoot on Sunday.

## GAME LAWS OF THE STATE OF NEW YORK.

§ 4. No person shall kill or expose for sale, or have in his possession after the same is killed, any wood duck (sometimes called summer duck), dusky duck (commonly called black duck), mallard, or teal duck, between the 1st of February and the 15th of August in each year, except on the waters of Long Island Sound or the Atlantic Ocean. No person shall at any time kill any wild duck, goose, or other wild fowl, with or by means of the device or instrument known as the swivel or punt gun, or with or by means of any gun other than such guns as are habitually raised at arm's length, and fired from the shoulder, or shall use any such device or instrument or gun other than such gun as aforesaid, with intent to kill any such duck, goose, or other wild fowl. No person shall in any manner kill, or molest with intent to kill, any wild ducks, geese, or other wild fowl, while the same are sitting at night upon their resting-places.

§ 5. Any person violating the foregoing provisions of this Act shall be deemed guilty of a misdemeanor, and shall likewise be liable to a penalty of fifty dollars for each offence; and it shall be the duty of all sheriffs, constables, and police officers to see that these provisions are enforced.

§ 6. No person shall at any time within this State, kill or trap, or expose for sale, or have in his possession after the same is killed, any eagle, fish hawk, night hawk, whip-poor-will, finch, sparrow, yellow bird, wren, martin, swallow, tonagar, oriole, bobolink, or any other song bird; or kill, trap, or expose for sale any robin, brown-thresher, wood-pecker, blackbird, meadow-lark, or starling, save during the months of August, September, October, November, and December; nor destroy or rob the nests of any wild birds whatever, under a penalty of five dollars for each bird so killed, trapped, or exposed for sale, and for each nest destroyed or robbed. This section shall not apply to any person who shall kill or trap any bird for the purpose of studying its habits or history, or having the same stuffed and set up as a specimen; nor to any person who shall kill on his own premises any robin during the period when summer fruits or grapes are ripening, providing such robin is killed in the act of destroying such fruits or grapes.

§ 7. No person shall at any time within ten years of the passage of this Act, kill any pinnated grouse, commonly called prairie-fowl, unless upon grounds owned by them, and grouse placed thereon by said owners, under a penalty of ten dollars for each bird so killed.

§ 8. No person shall kill, or have in his or her possession, except alive for the purpose of preserving the same alive through the winter, or expose for sale, any woodcock between the 1st of January and the 4th of July, or any quail, sometimes called Virginia partridge, between the 1st of January and the 20th of October, or any ruffed grouse, commonly called partridge, between the 1st of January and the 1st of September, or have in his possession any pinnated grouse, commonly called prairie-chicken, or expose the same for sale between the 1st of February and the 1st of July, under a penalty of ten dollars for each bird so killed or had in possession, or exposed for sale.

§ 11. There shall be no shooting, hunting, or trapping on the first day of the week, called Sunday; and any person violating the provisions of this section, shall be liable to a penalty of not more than twenty-five, nor less than ten dollars for each offence, or imprisonment for not more than twenty, nor less than five days.

§ 12. In the counties of Kings, Queens, and Suffolk, or on the waters adjacent to the same, no person shall kill, or have in his or her possession after the same is killed, any wild goose, brant, wood duck, dusky duck (commonly called black duck), mallard, widgeon, teal, sheldrake, broadbill, coot or old squaw, between the 10th of June and the 20th of October in each year; and no person shall kill or shoot at any wild goose, brant, or duck after sunset and before daylight on any day of the year; and no person shall sail for wild fowl or shoot at any wild goose, brant, or duck from any vessel propelled by sail or steam, or from any boat attached to the same; and no person shall use any floating battery or machine for the purpose of killing wild fowl, or shoot out of such floating machine at any wild goose, brant, or duck. But nothing herein contained shall prohibit the use of floats or batteries in Long Island Sound. Any person violating any of the provisions of this section shall be liable to a penalty of fifty dollars for each offence.

§ 13. Any person trespassing upon lands owned or occupied by another, for the purpose of shooting, hunting, or fishing thereon, after public notice by such owner or occupant as provided in the following section, shall be deemed guilty of trespass, and shall be liable to such owner or occupant in exemplary damages for each offence, not exceeding twenty-five dollars, and shall also be liable to the owner or occupant for the value of the gamo killed or taken.

[We may note here, in passing, some facts which show very clearly one phase of the misapprehension which exists among some of our readers about the names of common American birds and mammals. A writer may dislike to have names misapplied, yet his language will show he is ignorant of the zoological relations of the birds about which he writes. We may tell him that the birds which he, perhaps, calls *partridge, pinnated grouse, and grouse, are all of them grouse.* The *first is the ruffed*, the *second the pinnated*, and the *third the spruce grouse*, and any one of them may properly be called grouse. "*Bob White*" *is commonly called quail* in tho North, but throughout the South it is usually, and more correctly, called "*partridge*," which name in the New England States is invariably applied to the *ruffed grouse.* The ruffed grouse *is also called pheasant* in Pennsylvania, Minnesota, and the South, very incorrectly, of course. Strictly speaking, there are *no true quail or partridge indigenous to America*, but "Bob White" and his south-western cousins belong to the partridge family (*Perdicedæ*), and are so closely related to the true partridges that it is not a misuse of terms to give them that name.]

## Switzerland.

Throughout the Swiss Confederation game is universally recognized as the property of the State, but as each canton possesses sovereign rights within the narrow limits of its territory, the restrictive measures adopted for the preservation of game vary in many important respects. Notwithstanding the evident care with which these measures have been framed, and their gradually increasing stringency, they have not, however, been hitherto attended with any marked success, since the very existence of game, except perhaps in a few specially favoured localities, is generally admitted to be extremely problematical. How far this almost total disappearance of game of every kind may be attributed to the nature of the country itself, to the system of cultivation, to some inherent defect in the present law, or to laxity in administration, is a question not easily solved. Whatever the cause, the untoward effects are but too patent even to the most superficial observers. So much so that the greater number of the cantonal governments have within the last few years been empowered to use their discretion to the extent of either partially or wholly prohibiting the killing of game within their territories; but even measures of so exceptional a nature do not so far seem to have attained the object in view.

The system which obtains in all the cantons, with one or two exceptions, is that of requiring all persons engaged in killing game to provide themselves with special licences. These licences, available only for the period of one year, are made out in the names of the individuals to whom they have been granted and are not transferable. Any attempt to evade this regulation exposes both the holder of the licence and the person improperly using it to a heavy fine. They are issued by the Home Department only to such applicants as are either personally known to the department or recommended by the authorities of the district to which they belong.

Any one engaged in shooting game is bound to produce his licence at once when called upon to do so by police agents, forest guards, private keepers, and, in some cantons, any other licensed sportsman.

There are several kinds of licences, the charge for which differs in almost every canton. An ordinary licence, available for the whole of the shooting season, costs from six to twenty francs, but does not include large game ("hochgewild"), for which a special licence has to be obtained, costing in some cantons no less than forty francs. The fee charge for a licence merely to shoot snipe in the spring, or to be allowed to spread nets to catch birds of passage, varies in amount from four to ten francs. In some cantons, however, an extra charge is made when sporting dogs are used. In others, again, temporary permits are granted to non-residents for limited periods, at the rate of one franc fifty centimes per day. Lastly, the authorities in a few of the cantons are allowed to grant temporary permissions to distinguished foreigners, and to minors to go out in the company of regularly licensed sportsmen.

The only canton where game is said to be found in any considerable quantities is that of Aargau, which has not adopted the licence system. Its territory is divided for sporting purposes into seventy-two districts ("jagdreviere"), which are let on an eight years' lease by the State, at public auction, to the highest qualified bidders. The annual amount of the rent has to be paid in advance at the beginning of each year. The lessors cannot sublet, and not above six persons can enter into partnership to bid for the lease of a district. Their names have to be registered, and any subsequent changes among the co-lessors, should there be several, have to be at once notified to the proper authorities, and their consent thereto duly obtained. Leases cannot be held by individuals who are neither citizens of the canton nor have obtained the right of domicile.

The ordinary shooting season begins on the 1st of September, and generally ends on the 31st of December; but in some cantons it is not opened until the 1st of October. As a rule, shooting is not permitted in cultivated land until the crops have been completely gathered in, or in vineyards until the vintage is over. Moreover, the authorities reserve to themselves the right, under special circumstances, of delaying the period appointed for the commencement of the shooting season. Snipe, woodcocks, and other birds of passage may be killed between the beginning of March and the middle of April, and in some parts of the country up to the end of the latter month. In no instance does the season for this kind of sport extend beyond six weeks. Water fowl can be shot at all times, except between the 15th of April and the 1st of September. The season for killing deer, roe, and chamois extends from the beginning of September to the middle or end of October; but this description of game has of late become so exceedingly scarce that the great majority of the cantons have resolved to put a complete stop to its further destruction for a period of several years to come.

The sale of game out of the proper season, unless it can be proved to have been imported from abroad, is strictly forbidden, and both the vendor and buyer incur a heavy fine.

On Sundays and other holidays, shooting is likewise prohibited in every part of Switzerland.

Landed proprietors, farmers, and farm labourers may at any time destroy within the boundaries of their land, but without the aid of dogs; and either in any wood or public or private grazing-ground, all beasts and birds of prey, and destructive birds, game, or vermin, except hares. Polecats, martens, otters, foxes, wolves, lynxes, wild boars, bears, eagles, vultures, hawks, ravens, crows, magpies, sparrows, etc., are considered as vermin.

The destruction of singing birds, or such as are useful for agricultural purposes, as starlings, finches, titmice, larks, woodpeckers, etc., as well as their eggs and young, is a punishable offence.

Certain portions of the cantons where licences are granted are temporarily set apart as game preserves, and called "Yagdbaunbezirke." For a certain period, arbitrarily fixed by the authorities, no one is allowed to shoot in these districts.

No one is allowed to shoot near a house, within enclosures, or on fields and vineyards before the crop is removed.

## Persia.

I have ascertained that there are no laws for the preservation of game or for the prevention of trespass. The Shah, by virtue of his prerogative, can decree the strict preservation of the game in any portion of his dominions. In the neighbourhood of the capital itself, and in the immediate vicinity of the high-road, such a preserve exists, and so great is the terror entertained of the severe punishment his Majesty would probably inflict upon persons poaching on his grounds, that the most timid species of game are to be found within its limits, although, in the adjoining country, everything has been destroyed.

The Koran allows the pursuit of all kinds of game, with the exception of those that are forbidden or unclean. Acting upon this rule, every Persian who possesses a gun goes out and shoots where and when he pleases. He is liable to no penalty on account of trespass, save only where walled enclosures are concerned; and so strong is the feeling against any enclosure made solely for the preservation of game, that nothing of this nature is known to exist in Persia.

In short, game may be shot everywhere except on the Shah's preserves; nor can any one be prevented from trespassing, save in the case of a garden defended by high walls, and watched by specially appointed guardians. Where the corn is green and liable to be destroyed by the hoofs of horses, the owner may object; but if he be weaker than the trespasser, even in this case he has but the shadow of a chance of obtaining redress.

## Turkey.

No game laws have ever been framed by the Turkish Government; but there are certain police regulations which prohibit the killing of game at a certain season of the year,

These regulations, though tolerably well enforced in the neighbourhood of the capital, are generally little heeded by sportsmen in most of the provinces.

Game of every description is considered public property throughout the land, and may therefore be pursued and killed by anybody, provided he be furnished with a "teskéré," or licence, for carrying a gun, with which he must annually provide himself at the opening of the shooting season, beginning on the 1st of August and ending on the 31st of March.

In virtue of this "teskéré" all sportsmen acquire a right, already tacitly recognized, of shooting on any proprietor's land, as well as on crown lands. Shooting in the vicinity of the Sultan's kiosks, palaces, hospitals, barracks, and powder-magazines, is prohibited by the above-mentioned police regulations.

No laws of trespass exist, but the law forbids any person from entering a garden or field which may be surrounded by a stone wall.

Game is not preserved in Turkey.

## Denmark.

The particular birds and animals whose protection is a main cause of English game law discussions are seldom found here. It is said that there are no pheasants in Denmark, except in the king's preserves of Amack and Klampenborg; hares are very scarce, and rabbits are almost unknown.

Under these circumstances it would be useless to analyze the Danish game laws in detail.

The Danish "Vildt" has a wider meaning than our word "game." The law protects not only the nobler animals and birds which may be called "wild," but even such lower species as foxes, badgers, otters, martens, polecats, fieldfares, curlews, redshanks.

Creatures *feræ naturæ* are presumed to have no individual marks whereby they may be recognized, and they are held to belong to the land on which for the time being they are found. It follows that a sportsman cannot claim an animal or bird that escapes from his gun into another person's property.

Licences to carry arms, or for sporting, are not required.

Every one has a *primâ facie* right to deal as he pleases with shooting, trapping, or otherwise, or the game on his own land, be his tenure emphyteutical or freehold.

Any one may shoot wildfowl from a boat at sea, but a person so sporting may not wade along the shore, or shoot inwards on to the land, unless, of course, he is coasting his own estate.

There is little game in Denmark to tempt poachers; and the incidents of violence which follow poaching would be uncongenial to the quiet habits of the Danish peasant.

In order to favour the growth and settlement of dunes, where these are required for the protection of the coast, a so-called "peace," or jubilee, of several months is accorded by official order to certain sand-burrowing animals, such as foxes, martens, and the like, in the dunes named.

Deer and hares may not be killed between the 1st of March and 12th of September. For partridges, the fence period is 1st of February to 12th of September; for black-game and snipes, 1st of February to 1st of August.

Fines of five to ten dollars are inflicted on persons convicted of taking the nests or young of creatures classed as game.

Unauthorized persons of any kind taking singing birds, or injuring their nests, may be fined from two to five dollars.

No one may walk in another person's preserves with guns and dogs, unless lawful business call him, and his dogs are tied and his guns uncharged. Offenders against this rule are liable to be fined ten dollars, and to have their dogs shot by the competent proprietor or gamekeeper.

Field labourers taking loose dogs with them to their work are liable to a fine of ninepence for each offence. Excluded from this rule are shepherds' dogs and the like. The fine is not applicable where a dog follows its owner under circumstances such that no poaching intentions can be fairly assumed.

## PORTUGAL.

Every one is permitted to shoot, subject to the regulations imposed in the interests of agriculture.

In order to shoot game a licence is required, which is purchased from the Civil Government, the use of firearms being prohibited without such licence.

Every one may shoot on their own property, and on any cultivated lands (it seems) after the crop is gathered.

The municipality fixes annually the time when permission is allowed to sport on certain lands.

On open ground, planted with olive and other fruit trees, it is only during the period from the commencement of the fruit becoming ripe until gathered that such permission is withheld.

Game becomes the property of the sportsman on capturing it; he acquires a right to wounded game also.

If wounded game enters enclosed property the sportsman may not follow it, except with permission of the landowner. The sportsman can require the landowner, if present, to deliver up the dead game, or permit him to seek it.

Any damage done by the sportsman he is responsible for, if done in absence of landowner. If more than one sportsman, they are conjointly responsible.

Dogs entering enclosed property in pursuit of game, makes the sportsman responsible for damage done by them.

The owner of property, enclosed in such a manner as to

prevent easy egress or ingress of game, may shoot them whenever and in whatsoever manner he pleases.

Proprietors may destroy wild animals destructive to property.

It is strictly forbidden to destroy the eggs, or young game.

The municipalities determine the time when sporting is to cease altogether, and also what fines are to be imposed on persons who break the regulations.

## Spain.

Spaniards are forbidden to shoot or hunt on any ground which is not the private property of the individual, more especially in the provinces of Alasa, Avila, Burgos, Coruna, Guipuscoa, Huesca, Leon, Logrono, Lugo, Navarre, Orieuse, Oveido, Palencia, Pontevedra, Salamanca, Santander, Segovia, Soria, Valladolid, Viscaya, and Lamora, where all sport is forbidden from the 1st of April to the 1st of September; in the remaining provinces of Spain, including the Balearic and Canary Islands, all shooting is prohibited from the 1st of May to the 1st of August.

Another article of the Royal Decree prohibits shooting when the snow is lying upon the ground; and another one prohibits the use of traps, snares, nets, decoy birds, except for quails and birds of passage.

According to the 591st Article of the Penal Code, "people using firearms without a licence are punishable by a fine varying from five to twenty-five pesetas" (4*s.* to 20*s.*), and in the Article 608 of the same code it is stated that "persons trespassing on enclosed lands, or any private property, for the purpose of fishing or shooting, without permission from the proprietor, are amenable to the above-mentioned fine.

Two further articles of the same code state that the infringement of any laws for the protection of fish and game are punishable "by the same fine and the confiscation of the firearms or fishing implements of the delinquents."

These laws and ordinances are purely theoretical, and the practice of them is no longer observed in Spain. Shooting goes on at all times and seasons; snares, traps, and decoys are used all over the country; and the result is the most alarming decrease in every species of game throughout the country.

With respect to what game is property of the State and what of individuals, the law is as follows:—

All game in enclosed property, or property whose limits are defined and marked by large stones, stakes, or anything that is distinguishable, belongs to the proprietor of the soil, who can shoot it himself, let it, or give permission to his friends to shoot it.

The proper authorities, *i.e.* the governor of each province, can give permission to shoot on the lands belonging to the State or to the villages (communal lands), or on any private lands which are open: that is to say, which are not surrounded by walls or fences, and whose limits are not defined by landmarks, such as stones, posts, etc.

Any game alighting in private property, or falling wounded therein, belongs to the owner of the land, and not the parties who may have shot or hunted it.

Wolves, foxes, martens, wild cats, etc., are free game for all persons, and in all seasons.

## Grand Duchy of Baden.

The rights of killing and preserving game, which formerly belonged exclusively to the State and the feudal lords ("Standesherren" and "Grundherren"), were abolished as such in 1848, and transferred under certain conditions and regulations to the commune.

The communes hold a trust, not a right. They represent the landowners, and are compelled by the game law to let the shootings under their control by public auction for a period of at least three years.

Every landowner holding a compact estate of at least two hundred acres, is allowed the free independent exercise of his rights in regard to preserving and killing game.

*Regulations for protecting Game.*

Besides the special laws against poaching, the game law provides:—

1. That no game shall be killed or offered for sale between the 2nd of February and the 23rd of August, with the exception of wild boar, stags, roebucks, capercailzie, blackcocks, rabbits, and birds of passage.

2. An offence against the above is punishable by a fine of from five to twenty florins. Selling game out of season, stealing or wilfully destroying eggs or young birds, is punishable by a fine not exceeding ten florins.

From what has been already said, it will be seen that not only are the rights of landowners and sportsmen respected, as far as possible consistently with the public interests, but that the farmers are also protected against undue injury to their crops from over preserving. Wherever the head of game is proved to be excessive, the authorities may interfere and insist on its being reduced. There is consequently no ground in this country for regarding poaching as a venial offence, as if it were the natural result of arbitrary or oppressive laws. The actual degree of criminality to be attached to offences under the game laws is, nevertheless, viewed as a question not so easily determined, and as depending on various considerations.

Pursuing game with a gun on the land of others, without the knowledge or consent of the owner or his representatives ("Wilderei"), is punishable, according to circumstances, by imprisonment varying from fourteen days to four months.

The time when the offence was committed, whether day

or night, the character of the poacher, and the probable risk to which the property of the owner and the lives of the keepers or watchers would have been exposed, are all taken into account.

Snaring, or otherwise taking game without arms, is punishable by a fine of twenty-five to one hundred florins, to be paid to the owner. Repetition of the offence is punishable by eight days' to three months' imprisonment.

### *Property in Game.*

It is held in this country that no one can be admitted to possess the same perfect and equitable right of property in wild animals in a state of freedom, which he possesses in domesticated animals, or in game enclosed in parks or preserves, and thereby prevented from escaping. Consequently, game in a state of freedom is said to have no owner, and to belong to the State. The State, however, as before explained, concedes to the landowners, under certain conditions and limitations, the right of preserving and killing game on their estates, and declares by the game law that all game killed or found dead on their land is to be regarded as their property.

Game, on the other hand, which is enclosed, is exclusively the property of the landowner, or of any one duly qualified and authorized by him to occupy his place, whether that be the State or a private individual.

I have only to observe, in conclusion, that the game laws in Baden appear to work well, and to give general satisfaction. There is a large amount of game, especially roe, deer, and hares, in a state of freedom all over the valley of the Rhine in the Grand Duchy, a considerable part of which has hitherto gone to supply the Paris market.

Although this country is generally very fertile, and highly cultivated, few complaints about the game are heard, as far as I am aware, with the view to prove that the interests of agriculture are in danger. I can only attribute this to the general feeling of security arising from the fact that the authorities are at liberty to interfere, and do interfere, whenever in any district the game is found to have increased to an excessive degree.

### WURTEMBERG.

The law here distinctly shows that game is considered as the property of the individual and not of the State. But Wurtemberg being much broken up into small freehold properties, it would be impossible in practice to allow every one to shoot over his own plot; so, unless a man owns at least fifty acres, or that his bit of ground, if smaller, is properly fenced off, the parish, which is usually also a corporation, is owner of some of the woodland, lets the shooting of the smaller proprietors, for their benefit, with that belonging to the parish *en bloc.*

The shooting in Wurtemberg is not considered so good as in the neighbouring countries of Bavaria and Baden; the chief cause of this being attributed to the fact that the parishes, though they might let for a longer term if so minded, usually let their shootings by auction every three years; whereas, I am informed that, in Bavaria, six is the shortest term for a lease of shooting. The natural consequence of such a short lease as three years is, that the lessee not being sure of being able to secure the shooting again, shoots very hard the last season, and there is no time to get up a head of game between the different lettings.

Another clause considered by game preservers here as requiring alteration, is that which allows a man with fifty acres to retain the shooting thereof, as it can happen that a man may have a bit of bushy ground, a favourite resort of game in hard weather, and thus almost spoil a parish shooting district; and it is considered that it would be better if a minimum of from three hundred to four hundred acres were fixed, instead of fifty.

It may here be observed, that the land in most parts of Wurtemberg, is split up into small freeholds, the property of yeomen, who till these plots themselves, and who cannot be reimbursed, as most British farmers are where game is preserved, by holding their farms at a lower rent than they would have to pay were the game killed down and a re-valuation of the farm made; and the small amount of shooting-rent which the Wurtemberg yeoman might get on the division would not compensate him for any great damage. I am told, however, that no great damage is done by the game in Wurtemberg, though it sometimes happens that the roe are obliged to be shot pretty hard in some places where there are young plantations, as they eat off the tops of the young silver fir trees. The wild boars, which did most damage, have not existed outside game parks since 1848, when all game was destroyed to a great extent; nor have fallow deer, though I am told there is an attempt to get the latter up again in one district, and there are not many red deer, so that the number of the large game roaming about, and likely to do damage, is not excessive. As to the smaller game, there are occasionally complaints that hares bark the young fruit trees, of which there are great numbers in Wurtemberg; but this damage can be avoided if the trees are properly bound up, or smeared with a preparation; there are no wild rabbits, the soil not suiting them, and pheasants only in the royal pheasantry; so that the actual amount of damage done must be very small. Altogether, I am told that the game has greatly decreased since 1848, and that one hundred to one hundred and eighty hares are perhaps killed now, where four hundred to eight hundred were killed before that year.

As regards the fence months during the breeding seasons of the different kinds of game existing in Wurtemberg, a translation is annexed of the Royal Ordinance referred to in Article 12 of the Game Laws, as mentioned in that same Article 12. There exists another Ordinance protecting insectivorous birds useful to the garden, farm, and forest, and also singing birds, and prohibiting the taking of their eggs or young. In fact, small birds generally may be said to be protected, the Germans not only being generally by nature fond of and kind to them, but protecting them for the good they do in destroying the insects; starlings especially are protected, in many parts little boxes being put in trees in the cottage gardens for them to build in. For catching or shooting small birds a written permission from the "Oberamt" is necessary. Sparrows, if they become too numerous, are destroyed by a person specially authorized by the village to do so; so that the usual excuse of young men found creeping about lanes with a gun, "that they are only killing sparrows," cannot be given here. At the same time it may be remarked that small birds do not appear to be so numerous here as in most parts of England, probably from there not being any hedges, and also from the harder and longer frosts in winter killing off the weaker ones.

An actual law of trespass may be said not to exist in Wurtemberg; but when meadows are laid up for hay, and crops are standing in the fields, any one walking or riding across would be fined for the damage done. In the woods belonging to Government, people are also forbidden to quit the public paths, and persons gathering wild berries, etc., are obliged to be furnished with a permit. The parish woods could, I suppose, be closed in the same manner, but practically they are not.

Art. 10. The shooting licences are issued by the prefecture ("Oberamt"), and as a rule for natives of the country by the prefecture of the district in which the person desiring such licence resides, and for foreigners, from the prefecture of the district in which they intend to shoot. An appeal against the refusal to issue the same, can only be made to the court of the province ("Kreisbehörde") to which the "Oberamt" belongs.

Wild boars found outside game parks shall be destroyed as vermin.

Art. 13. Shooting is prohibited on holy days during the time of the morning service, and is forbidden entirely on Sundays and the great feast days.

Art. 16. Following the game is not allowed. The game which is wounded in another shooting district belongs to that person within the bounds of whose shooting it falls dead or is found.

§ 1. The fence months, during which game may neither be killed, trapped, exposed for sale, or bought, are fixed as follows, according to each different species of game :—

A. *Quadrupeds.*

1. Stags and bucks, 1st of October to 30th of June. Red and fallow deer.
2. Hinds (does), 1st of January to 30th of September. Red and fallow deer.
3. Roebucks, 1st of February to 31st of May.
4. Roe (does), 1st of January to 31st of October.
5. Hares, 1st of February to 31st of August.
6. Foxes, 1st of March to 30th of September.
7. Badgers, 1st of February to 31st of August.

B. *Feathered Game.*

1. Cock-of-the-wood (capercailzie) ("Auerhahn"), and blackcock, 16th of April to 31st of August.
2. Hazel-hens ("Haselhühner") (a sort of wood grouse), partridges, pheasants, from 1st of December to 31st of July.
3. Wild ducks, 1st of February to 31st of July.
4. Quails, wood pigeons, fieldfares, and thrushes, from 1st of March to 31st of August.

§ 2. Game, whether quadrupeds or feathered game, which is not included in § 1, can at any time of the year be killed, trapped, sold, and bought. Further, as to the prohibition to take away the eggs or young of feathered game, reference is to be made to Article 17, No. 9, of the law of the 27th October.

As concerns the protection of birds useful to the fields and forests, and singing birds, further directions will be given in a separate Ordinance.

§ 3. Whoever kills, traps, exposes for sale, or buys game during the fence months (§ 1) shall, in proportion to the magnitude of the offence, be punished by the "Oberamt," or the court of the province, with a fine not exceeding twenty-five florins (£2 2*s.*), according to Article 17, § 7, of the game law.

### Austro-Hungaria.

I wish only to observe that the views which guided the conception of Austrian laws had for object, on one hand, that the right of shooting should exclusively pertain to the owners of great estates, to persons enjoying the "droits seigneuriaux;" and, on the other hand, that the common peasants might be excluded from a sport, the indulgence in which might alienate them from their serious occupations. Thus in the course of years the principle made its way, but the right of shooting can only be a feudal right connected with the ownership of great landed properties.

1. The right of pursuing game on another man's property is abolished.

3. Villainage and other compulsory service for sporting purposes are abolished without indemnity.

5. Every proprietor of a rounded estate of at least two hundred jochs (one joch = about an acre) is entitled to pursue the game on his property.

6. On all other properties not excepted in §§ 4 and 5, situated within the limits of a commune, the game belongs, after the promulgation of this law, to the respective commune.

7. The commune is bound either to let the game without subdivision, or to exercise its right of pursuing it by means of trained gamekeepers.

8. The annual rent of the game thus ceded to the commune is, at the close of each year's agreement, to be divided amongst the proprietors, according to the extent of their property in the commune.

## Bavaria.

As regards the question of property in game (or more properly, in wild animals generally), the law of Bavaria recognizes no property in it, so long as it remains in its wild and natural state. Whilst in that condition it is neither the property of the State nor of individuals, but, as under the old Roman law, is held to be a *res nullius;* and it only becomes property after it has been reduced into possession, or acquired by legal means.

Down to the year 1848, the question of the right to kill game remained in Bavaria, as in most of the other German States, in very much the same condition as that which it had assumed three centuries previously.

The leading provisions of the law (of March 30, 1850) are to the following effect:—

It lays down the general principle that from and after the passing of this law the right to pursue or kill game shall be founded exclusively on the right of proprietorship in the land; that all previously existing seigniorial rights of the chase on land, the property of other persons, shall cease at once and for ever, and that no such rights shall ever again be created.

The general principle enunciated by the law as above stated, that the ownership of the land should, in future, constitute the foundation of the right to pursue or kill the game found upon it, is, however, at the same time, practically restricted, to a very notable extent, by the following proviso of the law: namely, that the exercise of this right shall be limited absolutely to the proprietors—whether nobles or peasants—of not less than two hundred and forty Bavarian acres * of land if situate in the plain, or of four hundred Bavarian acres if in the mountains; and further, to the proprietors of this extent of land only in those cases in which the entire two hundred and forty or four hundred acres respectively lie altogether so as to constitute one compact plot or parcel of land not intersected or divided by other lands. From this limitation, as to the rights of the chase, the law, however, exempts smaller portions of land if completely inclosed by a wall or other description of thick fence, as well as gardens and other plots of ground immediately attached to country houses or farm buildings, provided they be railed in; and further, no piece of land is held to be otherwise than compact or undivided in the sense of this law, if merely intersected by a road or stream.

As I have already stated, the number of proprietors of large landed estates, it may even be said of estates of more than a few hundred acres lying compactly together, is very limited in Bavaria, whilst by far the larger proportion of the land is in the hands of so-called peasant proprietors, owning on an average perhaps from fifty to a hundred acres.

Consequently, the practical result of the enactment above described is that, as a general rule, the rights of the chase in this country are enjoyed by persons who hire them from the communal authorities, the exception to the rule being the case of a proprietor exercising those rights on his own land.

The size of the communes varies very considerably in different districts of Bavaria; but I am informed that it may be assumed on an average, at about two thousand or three thousand Bavarian acres. The price usually obtained for the lease of the shooting depends, I need hardly state, very greatly on the locality of the commune, and on the quantity

* One hundred acres on "Tagwerke," are equal to 84½ English acres.

of game supposed to exist in it; but the rates are, as a rule, very low, especially if compared to what would be paid under similar circumstances in England. In the outlying rural districts, distant from any market towns, or where there may be but few resident proprietors inclined to hire the communal shootings, or where there is but little game, about fifty or one hundred florins * may be taken as the average annual rent of the shooting of a commune; whilst in the neighbourhood of large towns, or in localities where game is more than usually abundant, as much as six hundred florins is frequently paid.

It may be as well here to remark that a system of fences, or other mode of enclosing the land, is scarcely known in Bavaria; the boundary of each separate field or plot of ground being, as a general rule, marked by corner stones only, or by narrow paths.

In cases where a plot of land, consisting of less than the required two hundred and forty or four hundred acres, is completely surrounded by an extent of land belonging to one and the same person, sufficient to carry with it the rights of the chase, then the owner of the latter has the power of claiming the right to kill the game on the smaller piece of land so surrounded by his own, provided he pays to its owner an indemnity fixed according to the rates current for the hire of shootings in the district in which such land may be situated.

The chief descriptions of game found in Bavaria are (in the plains and cultivated lands generally), hares, and the common grey partridge; and in the woods and copses, with which this country is thickly studded, roe deer in considerable numbers, the latter being a kind of game which is much prized by German sportsmen. Indeed, from the German point of view, the roe deer constitute the leading and most attractive feature in the various elements of the chase in this country. Pheasants are rare, being only found in the

* Twelve Bavarian florins are equal to £1 sterling.

royal enclosed preserves, or "pheasantries," and in small numbers on some few private estates. In the large forests and mountainous districts, red deer are tolerably numerous; and there are, besides, capercailzies, or cock-of-the-wood, blackgame, ptarmigan, hazel-hens (a small description of wood grouse, unknown in the British Isles), and red-legged partridges; and the higher ridges on the Bavarian borders towards the Vorarlberg, the Tyrol, and Salzburg, afford some of the best chamois hunting in Europe. These last-named grounds belong chiefly to the crown, and are carefully preserved against intruders.

The following are some of the minor provisions of the law of the 30th of March, 1850, respecting the chase:—

No person is allowed to shoot, or otherwise go in pursuit of game, without being provided with a licence or card of permission. These cards are issued by the police authorities of the several districts to all persons applying for them, who are not under legal disability, and they are valid for one calendar year, and for the whole kingdom. Each card is available for one person only, and it must be made out in his name, and with his "signalement." The charge for each is eight florins. A fine, not exceeding twenty-five florins, to be recovered by the police authorities, or a proportionate term of imprisonment, is imposed upon all persons found in pursuit of game without being provided with one of the above-mentioned cards; or who make use of a card issued in the name of another person, or who take with them as a companion or guest in the chase a person carrying a gun and not provided with the necessary card; or who are found in pursuit of game on land (the right of shooting on which does not belong to them), without being accompanied by the person to whom that right belongs, or by an authorized guard or keeper; or who, whilst in pursuit of game, infringe the police regulations with reference to the protection of field crops, forests, etc.; and, lastly, who refuse to exhibit their card when called upon to do so by a duly authorized public officer. And, as regards persons found in pursuit of game without the necessary card, they incur, in addition to the above-mentioned fine of twenty-five florins, a further fine, equal in amount to the fee payable for the card itself.

Disturbing or taking the nests of capercailzies, black-game, hazel-hens, partridges, wild ducks or pheasants, or of any of the various kinds of the wild birds which breed in the fens or morasses.

All persons abetting or rendering assistance to others in the commission of infractions of this law are punishable by a fine, or term of imprisonment in proportion to the gravity of the offence.

The Royal Ordinance of October 5th, 1863, containing police regulations with reference to the chase, sets out by a declaration to the effect that the right of shooting or killing game shall in all cases be exercised with moderation, and with a due regard to the general interests of the chase, but at the same time forbids the maintenance of game in such quantities as to cause injury to the field crops or to the woods.

It then specifies the periods within which the killing of the different kinds of game and other wild animals is prohibited.

These periods are as follows:—

For stags (red deer), between the 15th of October and the 24th of June.

Hinds or yearlings, between the 6th of January and the 15th of September.

Fallow deer (bucks), between the 30th of October and the 24th of June.

Does, between the 6th of January and the 1st of October.

Chamois, between the 30th of November and the 25th of July.

Roebuck, between the 2nd of February and the 1st of June.

Wood hares, between the 2nd of February and the 15th of September.

Badgers, between the 1st of January and the 15th of September.

Beavers, between the 2nd of February and the 1st of October.

Marmots, between the 31st of October and the 15th of August.

Pheasants, between the 1st of March and the 1st of September.

Cock-of-the-wood and blackcock, between the 2nd of February and the 1st of August (except for a couple of weeks in April during the pairing season, when these birds may be shot).

Ptarmigan, hazel-hens, and red-legged partridges, between the 2nd of February and the 1st of August.

Wild ducks, between the 1st of March and the 30th of June.

Woodcocks and snipes, between the 15th of April (the 1st of May in the mountains) and the 1st of July.

Other birds which breed in the fens, and wild pigeons, fieldfares, etc., between the 1st of April and the 1st of June.

The further regulations laid down by this Ordinance are to the following effect. It is at all seasons of the year forbidden to shoot or take the does of roedeer, the young (less than a year old) of red deer, chamois, or roedeer, as also the hens of capercailzie and blackgame; but as regards the does of roedeer, if these should become so numerous in any locality as to appear to the person owning the right of shooting in such locality, to require thinning, special permission may be granted to him by the local police authorities, after consultation with the inspector of forests and chases of the district, to kill a certain number of them.

The opening of the shooting season for hares, partridges, and quails in the plain or open country, is fixed every year in each province of the kingdom, separately, by the chief provincial authority, on any day between the 15th of August and the 15th of September, according to the state of the harvest, and the day fixed upon must be announced in the official journal of the province. But any person having the right of shooting over an extent of land of not less than three thousand Bavarian acres lying altogether, may, on application to the police authorities of the district, obtain permission to shoot leverets for his own use at an earlier date than that which may be fixed upon by the authorities for the opening of the regular season for hare shooting.

The regulations above described, respecting the periods within which it is forbidden to kill the several kinds of game or other wild animals, only apply to the open country generally, and are not obligatory as regards the game kept in preserves completely enclosed by a paling or wall, or in pheasantries. They may, therefore, be considered a dead letter so far as fallow deer are concerned, none of these animals being found in Bavaria in a wild state in the forests (as in the neighbouring province of Bohemia), or otherwise than in enclosed preserves or parks, and even in the latter condition they are far from numerous.

The announcement of the opening of the shooting season confers no right to disregard the general prohibition with respect to walking over field crops still standing, or through vineyards in which the grapes have not been gathered; but this prohibition does not apply to pastures, clover, cabbages, potatoes, turnips, and mangel-wurzel.

It is expressly forbidden to shoot or otherwise take partridges so long as deep snow lies upon the ground.

Birds and beasts of prey may be shot or caught at all seasons of the year.

It is forbidden to make use of gun-cotton in shooting game, or to lay poisoned bait, or traps or nooses for the purpose of snaring game, except as regards birds of passage.

Red and fallow deer and chamois may not be shot otherwise than with the bullet.

The placing of spring-guns or man-traps is regulated by the provisions of Article 149 of the Criminal Police Code, which require that before such instruments be laid down the permission of the local authorities be obtained; that they only be laid down in grounds, woods, etc., which are completely enclosed by a paling or wall, and that a notice of their existence be affixed outside the enclosure.

It will, therefore, be readily understood, that the question is not one which can in this country cause much ill-feeling, or frequently give rise to litigation.

## FLORENCE.

Article 711 of the Italian Civil Code declares that property in game or fish ("gli animali che formano oggetto di caccia o di pesca") is acquired by occupancy.

Article 462 lays down the rule that pigeons, conies, and fish, passing from one pigeon-house, warren, or fish-pond into another, become the property of the owner of the latter, when they have not been artfully or fraudulently enticed.

By the communal and provincial law of 1865, every provincial council is empowered to determine the period during which the taking of game is to be permitted in each year.

## PIEDMONT.

Piedmontese Game Law of 1836, with modifications introduced by a law of 1853, extended to Lombardy by Decree July 29, 1859; to the Marches by Decree, November 21, 1860; and to Umbria, December 11, 1860.

It is not lawful to enter on another's land for the purpose of taking game, or to cause game to be hunted with dogs thereon, against the prohibition of the owner. Such prohibition shall always be presumed in the case of land sown or under crop, or enclosed with walls, hedges, or any other kind of fence, unless the owner's written permission to take game can be produced.

No one may shoot, or otherwise take game without a licence, which is personal and good for one year. The charge for a licence to shoot is ten francs, and thirty francs for a licence to take game with nets or snares, etc.

Offenders are liable to fines of eighty or one hundred and sixty francs, according to the degree and nature of the offence, when guns or dogs are used, and to fines of one hundred or two hundred francs when nets, etc., are employed. They may also be sent to gaol, for not less than eight days or more than one month in the former case; and, in the latter, for not more than two months or less than fifteen days.

A person trespassing on another's land, in pursuit of game, is further liable for any damage caused by him, and he must give up to the owner of the land all the game killed or taken thereon.

Any gun, net, dog, or other thing used in the taking of game, which the offender may have in his possession when found committing the offence, shall be immediately seized as security for the payment of fines or compensation.

The chase, at any time, of wolves, bears, and other animals, for the killing of which a reward is given, is likewise excepted. Such animals, however, must be hunted by soldiers belonging to Bersaglieri companies or to other arms, or by persons acting under the direction of the syndic of the commune.

## TUSCANY.

Decree of July 3, 1856.—The chase of animals and fowling are permitted to all persons.

No one may shoot who is not provided with a licence to carry arms.

Hunting and fowling on another's land, when it is not waste, without the owner's leave, are forbidden.

On waste land likewise they are forbidden, without such

consent, where the land is enclosed with walls, hedges, fences, or palings, and entirely surrounded by cultivated land, and if any permanent instrument or engines for fowling are employed.

The killing or taking of pigeons at any time and at any place is forbidden, under pain of a fine of thirty lire (£1) for every pigeon killed or taken. The aggregate amount of such fines, however, cannot exceed three hundred lire (£10). The birds are forfeited, as well as the arms or other instruments with which they are killed or taken.

It is forbidden to injure birds' nests and to take their eggs or nestlings, and likewise to injure the holes or lairs of wild four-footed animals, or to kill or take their young, any one of these offences being punishable with a fine of twenty lire (13*s.* 4*d.*); the aggregate amount of fine, however, not to exceed the sum of one hundred and fifty lire (£6).

From the above prohibition are at all times excepted young unfledged swallows, and the nests, eggs, nestlings of eagles, falcons, owls, ravens, jackdaws, magpies, sparrows, as well as the holes or dens and the young of wolves, foxes, polecats, martens, porcupines, hedgehogs, badgers, and weasels.

Any person who employs for the purpose of catching birds or other animals substances causing intoxication or stupefaction, and whosoever sets snares made of more than two horse hairs or of wire, and with which animals stronger than thrushes or blackbirds can be caught, shall pay a fine of from twenty to one hundred lire (13*s.* 4*d.* to £3 6*s.* 8*d.*).

All manner of hunting or fowling is prohibited when the ground is covered with snow, under pain of fine from twenty to one hundred lire, together with forfeiture of arms or instruments employed.

The pursuit of game, etc., with a gun, from one hour after sunset until one hour before sunrise, is prohibited under pain of fine from thirty to one hundred lire (£1 to £3 6*s.* 8*d.*).

This prohibition, however, is not applicable to shooting in marshes. Any one carrying a loaded gun between the hours aforesaid, in going to or returning from his shooting-ground, incurs the same penalty.

The penalty for hunting or fowling in any manner at a time when they are not permitted is a fine not exceeding one hundred and fifty lire, or less than fifty lire, together with the forfeiture of guns or other instruments. To the same fine is liable whosoever during the time above mentioned lays snares for any kind of animals, or does not remove such snares, etc., previously laid by him, or carries a gun on a public road or in the open country, or carries any implements or engines used in fowling, or transports, deals in, or keeps in his possession game of any kind.

It is lawful, however, to hunt or take at such time noxious beasts and birds, such as wolves, foxes, badgers, polecats, martens, weasels, porcupines, hedgehogs, falcons, owls, ravens, jackdaws, magpies, and crows; provided that, in so doing, neither guns, nor snares, nor traps are used; and sparrows may be caught by any means, but they may not be shot with guns.

Prefects may, during the season of general prohibition, give permission for a determined number of days to companies comprising not fewer than eight persons, to shoot wolves and foxes with guns. In certain particular cases they may grant such permission to less than eight individuals together, and they may at any time permit the use of snares and traps even in fields, woods, and other open places where it is necessary to employ such means of protection against the animals above mentioned, provided that such traps, etc., be set an hour after sunset, and removed an hour before sunrise, and that they be not set in roads, paths, or tracks where men or animals pass. Such permission may be made applicable to wild boars when their increase becomes injurious to agriculture.

From the close of the shooting season until the 16th of

March, wood pigeons may be shot by special permission in places where it is customary to do so.

Special permission may also be obtained to shoot water-fowl until the 14th of April, on lakes, marshes, and ponds, on the Arno, the Serchio, the Chiara Canal from the Lake of Chiusi to the Arno, the Tiber from Piene S. Stefano to the Pontifical frontier, the Ombrone from the confluence of the Arbia to the sea, and on the Cecina from the confluence of the Possera to the sea.

During the period above specified, woodcock shooting is permitted only on lakes, marshes, and ponds.

During the same period, ending 14th of April, it is lawful for any person to catch lapwings, plovers, starlings, and "gambette;" but the use of lime-twigs, traps, or nets with close meshes, is forbidden, under pain of fine, from forty to one hundred lire.

### Tuscany.

From and after the 8th of August, quails, turtle doves, fig-eaters ("beccafichi"), ortolans, nightingales, gulls (?), and other small birds which leave Tuscany in the course of the summer, may be caught with open nets, lime twigs, and in other ways specified in the law.

Quails may be shot by special permission from the 16th to the 31st of August.

### Neapolitan Provinces.

The regulations in force are mainly founded on a decree of Ferdinand I., dated 18th of August, 1819. They are to the following effect:—

No one may shoot, or go in pursuit of game at any season, or in any place, without a licence, under pain of forfeiture of gun, etc., and a fine of fifty ducats (about £8 10*s.*), besides the punishment awarded by the penal laws for carrying arms without permission. Formerly two licences were required: a licence to carry arms, and a licence to shoot game. One licence is now sufficient, called, "Permesso di Armi e di Caccia:" this licence is obtained from the head of the police department, in the chief town of each province, who delivers it at his discretion to persons of whose respectability he is assured. The charge for it, which varies in different provinces, is at Naples equivalent to about 11*s.* 2*d.*

Any person provided with such a licence, may shoot in the open country; but it is forbidden to go in pursuit of game into royal preserves, or upon any grounds enclosed with walls, hedges, ditches, or banks of earth of the height of four feet four inches, without the owner's leave, under pain of forfeiture of gun, accoutrements, etc., together with a fine not exceeding ten ducats (about £1 14*s.*). The same prohibition extends to unenclosed vineyards from the 1st of September to the close of the vintage.

A similar penalty is incurred by any person shooting, or going in pursuit of game, from the 1st of April to the 30th of August.

Quails, and other birds of passage, however, may be taken or shot on the sea-shore, in the months of April and May, and on uncultivated ground elsewhere, in June and July.

The employment of snares, or nooses to catch hares, partridges, woodcocks, or pheasants, is prohibited at all times, and in all places, under penalty of a fine of ten ducats (about £1 14*s.*), and imprisonment for a term not exceeding fifteen days. The same penalty is incurred by any person shooting another's pigeons in the fields, taking eggs from the nests of quails, partridges, pheasants, and blackcocks, or taking the young of hares or deers.

The above-mentioned penalties may be doubled in the case of offences committed during the night.

## Sicily.

I am informed that any person provided with a licence is at liberty to shoot game of any description, and at all seasons wherever he finds it, except within walled enclosures, which are devoted to the preservation of game for the use of the proprietor of the land. In every case of trespass, whether committed in pursuit of game or otherwise, the proprietor's remedy is to lodge a complaint against the offender with the local judicial authorities, by whom the trespasser, if convicted, may be sentenced to make good any actual damage, and to pay a small fine, besides costs.

## Venetian Provinces.

No one can be authorized to use poison, to hunt or shoot, etc., hares when the ground is covered with snow, to hunt stags, fallow deer, or roebucks, to hunt with hounds in the fields before the end of September, or to go in pursuit of game, etc., on another person's land which is enclosed, or, if unenclosed, on which there are any kinds of produce liable to damage. The penalty for the commission of any of these offences is a fine of one hundred and eighty francs.

Land is considered as enclosed only when it is completely surrounded by fences or ditches in such a manner as to show manifestly the intention of the owner constantly to prevent the ingress of persons as well as beasts.

A licence is only valid from the 1st of July to the 15th of the following April. Shooting, fowling, etc., at any other time are punishable with fines of one hundred and eighty francs.

## Roman Province.

A law of August 14th, 1839, declares that all persons may

chase both quadrupeds and birds under the following regulations:—

From the 1st of April to the 1st of August, the chase of useful quadrupeds or birds, with the exception of quails, which may be taken on the sea-shore, but not elsewhere, at the time of their arrival, is prohibited.

During the same time no one is allowed to sell or buy game of any sort, except quails at the time of their arrival.

The spoiling of eggs or nests, and the killing of the young of useful animals are prohibited. It is also forbidden to pursue hares, roebucks, partridges, and other useful birds or quadrupeds in places covered with snow.

No one may at any time take or kill pigeons, the property of another.

Without the owner's leave, no one may go in pursuit of game on another person's land, if it be enclosed with walls, hedges, or other fences in such a manner as to prevent the entrance of both men and beasts, or, even if not so enclosed, when it is under crop or prepared for cultivation. This provision is applicable to unenclosed property in marshy districts yielding natural produce of various kinds.

## ITALY.

The following are the duties chargeable on game licences in different parts of the kingdom of Italy:—

| | Shooting Licence. | Licence to take Game with nets, etc. |
|---|---|---|
| | *Fr. c.* | *Fr. c.* |
| Province of the former kingdom of Sardinia, Lombardy, Romagna, and the Marches ... .. ... | 10 0 | 30 0 |
| Pemgia ... ... ... ... ... | 10 0 | 18 40 |
| Massa Carrara ... ... ... ... | 10 30 | 30 0 |
| Modena ... ... ... ... | 9 50 | 30 0 |

| | Shooting Licence. | Licence to take Game with nets, etc. |
|---|---|---|
| Reggio-Emilia ... ... ... ... | 10 0 | 30 0 |
| Piacenza, Parma ... ... ... ... | 12 0 | 6 0 |
| Tuscany ... ... ... ... | 13 40 | ... |
| Naples ... ... ... ... ... | 12 75 | Various: from 2 12 to 6 37 |
| Neapolitan Provinces... ... ... | 8 50 | From 2 12 to 4 25 |
| Sicily: | | Various: from 1 6 to 12 75 |
| Licence to shoot game ... ... | 6 37 | |
| Licence to bear arms ... ... | 10 0 | |

To these duties must be added the war tenth and stamps, amounting to one franc twenty centimes.

## Saxony.

According to the laws of Saxony, the right of killing game extends to all those animals and birds (living in their natural condition of freedom, and therefore constituting public property) which have hitherto been considered as game in this country, viz. red deer, fallow deer, roedeer, wild boar, wild rabbits, hares, beavers, badgers, otters, foxes, martens, fitchets, weasels, ermines, wild cats, squirrels, and all wild birds.

He who has the right of killing game is also entitled to destroy the nests of wild birds in his district, to take out their eggs and young ones, and to take possession of dying game and of shed stag horns.

The owners of the grounds on which the "Altberechtigten" have the right of killing game are entitled to redeem this privilege by paying an indemnity for it, the amount of which is fixed by law.

The right of killing game belongs, furthermore, to all proprietors and usufructuaries of estates, which form an unintersected area of at least three hundred acres of field or woodland.

Railway roads and rivers are not to be considered as intersecting a hunting district, with the single exception of the River Elbe.

The owners of smaller estates have to form conjoint-hunting districts with their neighbours, which must at least extend over three hundred acres.

In such hunting districts the right of killing game cannot be exercised by single proprietors of the grounds of which the aforesaid districts are composed, but only by foresters duly appointed, or by persons who have rented the right of killing game in the districts in question.

Even the persons who possess the right of killing game are not allowed to make use of this right throughout the whole year.

A time has, on the contrary, been fixed, during which it is forbidden to kill game. This time extends—

1. For red deer and fallow deer, from the 1st of April to the 15th of July inclusively.

2. For wild ducks, from the 1st of April to the 15th of June inclusively.

3. For all other game, from the 1st of February to the 31st of August.

Persons killing game during this time are fined or imprisoned. This law does not, however, apply to the killing of beasts of prey, such as otters, foxes, martens, fitchets, weasels, wild cats, etc.

Moreover, it is forbidden—

1. To hunt or shoot game in premises and places which are inhabited.

2. To make use of cruel means for hunting or shooting game.

3. To kill game on Sundays during the time of divine service, and in the neighbourhood of churches and cemeteries.

Driving game is entirely forbidden on Sundays.

Lastly, it is to be remarked that the pursuit of wounded game into another person's hunting district is forbidden.

### Belgium.

The regulations relative to game and to the law of trespass in pursuit of game in Belgium are governed by the law of the 26th of February, 1846. The *Moniteur Belge*, in the "Exposé des Motifs" for this law (p. 1227), says: "The Constituent Assembly, in destroying the feudal *regimé*, has, by its decree of the 4th–11th of August, 1789, considered the right of shooting or destroying game as inherent to the land. The execution of this decree having given rise to grave disorders which it was necessary to repress in the interest of agriculture, the same Assembly, by the law of the 28th–30th of April, 1790, fixed certain limits to the right of pursuing game."

The material points of this law are—

1. The fixing by the Government of the periods for opening and closing the right to shoot or otherwise pursue game.

2. Prohibition of every description of pursuing game, either with gun, by coursing, by nets or snares, out of these periods.

3. Prohibition to pursue game over another person's land without the consent of the proprietor, or of the person holding a right from him.

4. Prohibition to remove or destroy on another's land eggs or broods of quails, pheasants, partridges, blackcock, rails, grouse, plover, and waterfowl.

Absolute prohibition, in or out of the stated periods, of snares, nets, baited and other traps suitable for taking or destroying pheasants, partridges, quails, blackcock, gelinottes, rails, grouse, plovers, snipe, jacksnipe, hares, rabbits, chévreuil, stags, or deer.

Thus, as to the game mentioned above, no sport can take place but by shooting or coursing, but rabbits can at all times be taken with nets and ferrets.

6. Absolute prohibition, after the closing of the season, of using nets, snares, or engines applicable to or capable of taking or destroying any sort of game not herein specified.

7. Prohibition to expose for sale, buy, or hawk, during the close season, quails, pheasants, partridges, gelinottes, blackcock, rails, snipe, jacksnipe, hares, roebucks, stags, and deer.

Article 2 of the law above referred to reproduces the ancient legislation and the principles of the law of 1790 as to the ownership of the right of pursuit of game. Every kind of right, even in the matter of small birds, is forbidden on the land of another without the proprietor's consent. The right to game is a right inherent to the property. The possessor of the soil has, therefore, the right to dispose of it. He may transmit this right to a third person, that is to say, he may let or cede the game on his property. In that case this third party is the representative of the owner. The farmer to whom the right of game has not been granted under his lease cannot sport without the permission of the landlord.

Poaching prevails largely in Belgium, especially in the vicinity of large manufacturing towns, many of the workmen in which, preferring a life of crime to the pursuit of an honest calling, organize themselves in bands more or less numerous, and systematically endeavour to enrich themselves at the expense of their neighbours.

THE END.

PRINTED BY WILLIAM CLOWES AND SONS, LIMITED, LONDON AND BECCLES.

# *A LIST OF BOOKS*

PUBLISHED BY

# CHATTO & WINDUS,

## 214, PICCADILLY, LONDON, W.

*Sold by all Booksellers, or sent post-free for the published price by the Publishers.*

---

**About.—The Fellah**: An Egyptian Novel. By EDMOND ABOUT. Translated by Sir RANDAL ROBERTS. Post 8vo, illustrated boards, 2s.; cloth limp, 2s. 6d.

---

**Adams (W. Davenport), Works by:**

**A Dictionary of the Drama.** Being a comprehensive Guide to the Plays, Playwrights, Players, and Playhouses of the United Kingdom and America, from the Earliest to the Present Times. Crown 8vo, half-bound, 12s. 6d. [*Preparing.*

**Latter-Day Lyrics.** Edited by W. DAVENPORT ADAMS. Post 8vo, cloth limp, 2s. 6d.

**Quips and Quiddities.** Selected by W. DAVENPORT ADAMS. Post 8vo, cloth limp, 2s. 6d.

---

**Advertising, A History of,** from the Earliest Times. Illustrated by Anecdotes, Curious Specimens, and Notices of Successful Advertisers. By HENRY SAMPSON. Crown 8vo, with Coloured Frontispiece and Illustrations, cloth gilt, 7s. 6d.

---

**Agony Column (The) of "The Times,"** from 1800 to 1870. Edited, with an Introduction, by ALICE CLAY. Post 8vo, cloth limp, 2s. 6d.

---

**Aidé (Hamilton), Works by:**

Post 8vo, illustrated boards, 2s. each.

**Carr of Carrlyon.**

**Confidences.**

---

**Alexander (Mrs.), Novels by:**

**Maid, Wife, or Widow?** Crown 8vo, cloth extra, 3s 6d.; post 8vo, illustrated boards, 2s.

**Valerie's Fate.** Post 8vo, illust. bds., 2s.

---

**Allen (Grant), Works by:**

Crown 8vo, cloth extra, 6s. each.

**The Evolutionist at Large.** Second Edition, revised.

**Vignettes from Nature.**

**Colin Clout's Calendar.**

**Strange Stories.** With Frontispiece by GEORGE DU MAURIER. Cr. 8vo, cl. ex., 6s.; post 8vo, illust. bds., 2s.

**Philistia:** A Novel. Crown 8vo, cloth extra, 3s. 6d.; post 8vo, illust. bds., 2s.

**Babylon:** A Novel. With 12 Illusts. by P. MACNAB. Crown 8vo, cloth extra, 3s. 6d.

**For Maimie's Sake:** A Tale of Love and Dynamite. Cr. 8vo, cl. ex., 6s.

**In all Shades:** A Novel. Three Vols., crown 8vo. [*Shortly.*

---

**Architectural Styles, A Handbook of.** Translated from the German of A. ROSENGARTEN, by W. COLLETT-SANDARS. Crown 8vo, cloth extra, with 639 Illustrations, 7s. 6d.

---

**Artemus Ward:**

**Artemus Ward's Works:** The Works of CHARLES FARRER BROWNE, better known as ARTEMUS WARD. With Portrait and Facsimile. Crown 8vo, cloth extra, 7s. 6d.

**Artemus Ward's Lecture on the Mormons.** With 32 Illustrations. Edited, with Preface, by EDWARD P. HINGSTON. Crown 8vo, 6d.

**The Genial Showman:** Life and Adventures of Artemus Ward. By EDWARD P. HINGSTON. With a Frontispiece. Cr. 8vo, cl. extra, 3s. 6d.

---

**Art (The) of Amusing:** A Collection of Graceful Arts, Games, Tricks, Puzzles, and Charades. By FRANK BELLEW. With 300 Illustrations. Cr. 8vo, cloth extra, 4s. 6d.

**Ashton (John), Works by:**
Crown 8vo, cloth extra, 7s. 6d. each.
**A History of the Chap-Books of the Eighteenth Century.** With nearly 400 Illustrations, engraved in facsimile of the originals.
**Social Life in the Reign of Queen Anne.** From Original Sources. With nearly 100 Illustrations.
**Humour, Wit, and Satire of the Seventeenth Century.** With nearly 100 Illustrations.

---

**English Caricature and Satire on Napoleon the First.** With 120 Illustrations from Originals. Two Vols., demy 8vo, cloth extra, 28s.

**Bacteria.—A Synopsis of the Bacteria and Yeast Fungi and Allied Species.** By W. B. GROVE, B.A. With 87 Illusts. Crown 8vo, cl. extra, 3s. 6d.

**Bankers, A Handbook of London;** together with Lists of Bankers from 1677. By F. G. HILTON PRICE. Crown 8vo, cloth extra, 7s. 6d.

**Bardsley (Rev. C.W.), Works by:**
Crown 8vo., cloth extra, 7s. 6d. each.
**English Surnames:** Their Sources and Significations. Third Ed., revised.
**Curiosities of Puritan Nomenclature.**

**Bartholomew Fair, Memoirs of.** By HENRY MORLEY. With 100 Illusts. Crown 8vo, cloth extra, 7s. 6d.

**Beaconsfield, Lord: A Biography.** By T. P. O'CONNOR, M.P. Sixth Edition, with a New Preface. Crown 8vo, cloth extra, 7s. 6d.

**Beauchamp. — Grantley Grange:** A Novel. By SHELSLEY BEAUCHAMP. Post 8vo, illust. bds., 2s.

**Beautiful Pictures by British Artists:** A Gathering of Favourites from our Picture Galleries. In Two Series. All engraved on Steel in the highest style of Art. Edited, with Notices of the Artists, by SYDNEY ARMYTAGE, M.A. Imperial 4to, cloth extra, gilt and gilt edges, 21s. per Vol.

**Bechstein. — As Pretty as Seven,** and other German Stories. Collected by LUDWIG BECHSTEIN. With Additional Tales by the Brothers GRIMM, and 100 Illusts. by RICHTER. Small 4to, green and gold, 6s. 6d.; gilt edges, 7s. 6d.

**Beerbohm. — Wanderings in Patagonia;** or, Life among the Ostrich Hunters. By JULIUS BEERBOHM. With Illusts. Crown 8vo, cloth extra, 3s. 6d.

**Belgravia for 1886. — One Shilling Monthly.** Illustrated by P. MACNAB.—The first Chapters of **Mohawks**, a New Novel by M. E. BRADDON, Author of "Lady Audley's Secret," appeared in the JANUARY Number, and the Story will be continued throughout the year. This Number contained also the Opening Chapters of a New Novel entitled **That other Person;** and several of those short stories for which *Belgravia* is famous.
*** *Now ready, the Volume for* MARCH *to* JUNE 1886, *cloth extra, gilt edges,* 7s. 6d.; *Cases for binding Vols.*, 2s. *each.*

**Belgravia Annual for Christmas, 1886.** Demy 8vo, with Illustrations, 1s. [*Preparing.*

**Bennett (W.C., LL.D.), Works by:**
Post 8vo, cloth limp, 2s. each.
**A Ballad History of England**
**Songs for Sailors.**

**Besant (Walter) and James Rice, Novels by.** Crown 8vo, cloth extra, 3s. 6d. each; post 8vo, illust. boards, 2s. each; cloth limp, 2s. 6d. each.

**Ready-Money Mortiboy.**
**With Harp and Crown.**
**This Son of Vulcan.**
**My Little Girl.**
**The Case of Mr. Lucraft.**
**The Golden Butterfly.**
**By Celia's Arbour.**
**The Monks of Thelema.**
**'Twas in Trafalgar's Bay.**
**The Seamy Side.**
**The Ten Years' Tenant.**
**The Chaplain of the Fleet.**

**Besant (Walter), Novels by:**
Crown 8vo, cloth extra, 3s. 6d. each post 8vo, illust. boards, 2s. each; cloth limp, 2s. 6d. each.
**All Sorts and Conditions of Men:** An Impossible Story. With Illustrations by FRED. BARNARD.
**The Captains' Room, &c.** With Frontispiece by E. J. WHEELER.
**All in a Garden Fair.** With 6 Illusts, By H. FURNISS.
**Dorothy Forster.** With Frontispiece By CHARLES GREEN.
**Uncle Jack, and other Stories.**

**Children of Gibeon:** A Novel. Three Vols., crown 8vo. [*Shortly.*
**The Art of Fiction** Demy 8vo, 1s.

**Betham-Edwards (M.), Novels** by. Crown 8vo, cloth extra, 3s. 6d. each.; post 8vo, illust. bds., 2s. each.

Felicia. | Kitty.

**Bewick (Thos.) and his Pupils.** By AUSTIN DOBSON. With 95 Illustrations. Square 8vo, cloth extra, 10s. 6d.

**Birthday Books:—**

The Starry Heavens: A Poetical Birthday Book. Square 8vo, handsomely bound in cloth, 2s. 6d.

Birthday Flowers: Their Language and Legends. By W. J. GORDON. Beautifully Illustrated in Colours by VIOLA BOUGHTON. In illuminated cover, crown 4to, 6s.

The Lowell Birthday Book. With Illusts. Small 8vo, cloth extra, 4s. 6d.

**Blackburn's (Henry) Art Handbooks.** Demy 8vo, Illustrated, uniform in size for binding.

Academy Notes, separate years, from 1875 to 1885, each 1s.

Academy Notes, 1886. With numerous Illustrations. 1s.

Academy Notes, 1875-79. Complete in One Vol., with nearly 600 Illusts. in Facsimile. Demy 8vo, cloth limp, 6s.

Academy Notes, 1880-84. Complete n One Volume, with about 700 Facsimile Illustrations. Cloth limp, 6s.

Grosvenor Notes, 1877. 6d.

Grosvenor Notes, separate years, from 1878 to 1885, each 1s.

Grosvenor Notes, 1886. With numerous Illustrations. 1s.

Grosvenor Notes, 1877-82. With upwards of 300 Illustrations. Demy 8vo, cloth limp, 6s.

Pictures at South Kensington. With 70 Illusts. 1s. [*New Edit. preparing.*

The English Pictures at the National Gallery. 114 Illustrations. 1s.

The Old Masters at the National Gallery. 128 Illustrations. 1s. 6d.

A Complete Illustrated Catalogue to the National Gallery. With Notes by H. BLACKBURN, and 242 Illusts. Demy 8vo, cloth limp, 3s.

---

Illustrated Catalogue of the Luxembourg Gallery. Containing about 250 Reproductions after the Original Drawings of the Artists. Edited by F. G. DUMAS. Demy 8vo, 3s. 6d.

The Paris Salon, 1885. With about 300 Facsimile Sketches. Edited by F. G. DUMAS. Demy 8vo, 3s.

ART HANDBOOKS, *continued*—

The Paris Salon, 1886. With about 300 Illusts. Edited by F. G. DUMAS. Demy 8vo, 3s.

The Art Annual, 1883-4. Edited by F. G. DUMAS. With 300 full-page Illustrations. Demy 8vo, 5s.

**Blake (William): Etchings from** his Works. By W. B. SCOTT. With descriptive Text. Folio, half-bound boards, India Proofs, 21s.

**Boccaccio's Decameron; or,** Ten Days' Entertainment. Translated into English, with an Introduction by THOMAS WRIGHT, F.S.A. With Portrait, and STOTHARD'S beautiful Copper-plates. Cr. 8vo, cloth extra, gilt, 7s. 6d.

**Bowers'(G.) Hunting Sketches:** Oblong 4to, half-bound boards, 21s. each.

Canters in Crampshire.

Leaves from a Hunting Journal. Coloured in facsimile of the originals.

**Boyle (Frederick), Works by:** Crown 8vo, cloth extra, 3s. 6d. each; post 8vo, illustrated boards, 2s. each.

Camp Notes: Stories of Sport and Adventure in Asia, Africa, and America.

Savage Life: Adventures of a Globe-Trotter.

Chronicles of No-Man's Land. Post 8vo, illust. boards, 2s.

**Braddon (M. E.)—Mohawks,** a Novel, by Miss BRADDON, Author of "Lady Audley's Secret," was begun in BELGRAVIA for JANUARY, and will be continued throughout the year. Illustrated by P. MACNAB. 1s. Monthly.

**Brand's Observations on Popular Antiquities,** chiefly Illustrating the Origin of our Vulgar Customs, Ceremonies, and Superstitions. With the Additions of Sir HENRY ELLIS. Crown 8vo, cloth extra, gilt, with numerous Illustrations, 7s. 6d.

**Bret Harte, Works by:**

Bret Harte's Collected Works. Arranged and Revised by the Author. Complete in Five Vols., crown 8vo, cloth extra, 6s. each.

Vol. I. COMPLETE POETICAL AND DRAMATIC WORKS. With Steel Portrait, and Introduction by Author.

Vol. II. EARLIER PAPERS—LUCK OF ROARING CAMP, and other Sketches—BOHEMIAN PAPERS — SPANISH AND AMERICAN LEGENDS.

Vol. III. TALES OF THE ARGONAUTS —EASTERN SKETCHES.

Vol. IV. GABRIEL CONROY.

Vol. V. STORIES — CONDENSED NOVELS, &c.

BRET HARTE, *continued*—

**The Select Works of Bret Harte**, in Prose and Poetry. With Introductory Essay by J. M. BELLEW, Portrait of the Author, and 50 Illustrations. Crown 8vo, cloth extra, 7s. 6d.

**Bret Harte's Complete Poetical Works.** Author's Copyright Edition. Beautifully printed on hand-made paper and bound in buckram. Cr. 8vo, 4s. 6d.

**Gabriel Conroy:** A Novel. Post 8vo, illustrated boards, 2s.

**An Heiress of Red Dog,** and other Stories. Post 8vo, illustrated boards, 2s.

**The Twins of Table Mountain.** Fcap. 8vo, picture cover, 1s.

**Luck of Roaring Camp,** and other Sketches. Post 8vo, illust. bds., 2s.

**Jeff Briggs's Love Story.** Fcap. 8vo, picture cover, 1s.

**Flip.** Post 8vo, illustrated boards, 2s.; cloth limp, 2s. 6d.

**Californian Stories** (including THE TWINS OF TABLE MOUNTAIN, JEFF BRIGGS'S LOVE STORY, &c.) Post 8vo, illustrated boards, 2s.

**Maruja:** A Novel. Post 8vo, illust. boards, 2s.; cloth limp, 2s. 6d.

**The Queen of the Pirate Isle.** With 25 original Drawings by KATE GREENAWAY, Reproduced in Colours by EDMUND EVANS. Small 4to, boards, 5s. [*Shortly.*

## Brewer (Rev. Dr.), Works by:

**The Reader's Handbook of Allusions, References, Plots, and Stories.** Fifth Edition, revised throughout, with a New Appendix, containing a COMPLETE ENGLISH BIBLIOGRAPHY. Cr. 8vo, 1,400 pp., cloth extra, 7s. 6d.

**Authors and their Works, with the Dates:** Being the Appendices to "The Reader's Handbook," separately printed. Cr. 8vo, cloth limp, 2s.

**A Dictionary of Miracles:** Imitative, Realistic, and Dogmatic. Crown 8vo, cloth extra, 7s. 6d.; half-bound, 9s.

## Brewster (Sir David), Works by:

**More Worlds than One:** The Creed of the Philosopher and the Hope of the Christian. With Plates. Post 8vo, cloth extra 4s. 6d.

**The Martyrs of Science:** Lives of GALILEO, TYCHO BRAHE, and KEPLER. With Portraits. Post 8vo, cloth extra, 4s. 6d.

**Letters on Natural Magic.** A New Edition, with numerous Illustrations, and Chapters on the Being and Faculties of Man, and Additional Phenomena of Natural Magic, by J. A. SMITH. Post 8vo, cl. ex., 4s. 6d.

## Briggs, Memoir of Gen. John.

By Major EVANS BELL. With a Portrait. Royal 8vo, cloth extra, 7s. 6d.

## Brillat-Savarin.—Gastronomy

**as a Fine Art.** By BRILLAT-SAVARIN. Translated by R. E. ANDERSON, M.A. Post 8vo, cloth limp, 2s. 6d.

## Buchanan's (Robert) Works:

Crown 8vo, cloth extra, 6s. each.

**Ballads of Life, Love, and Humour.** Frontispiece by ARTHUR HUGHES.

**Undertones.**

**London Poems.**

**The Book of Orm.**

**White Rose and Red:** A Love Story.

**Idylls and Legends of Inverburn.**

**Selected Poems of Robert Buchanan.** With a Frontispiece by T. DALZIEL.

**The Hebrid Isles:** Wanderings in the Land of Lorne and the Outer Hebrides. With Frontispiece by WILLIAM SMALL.

**A Poet's Sketch-Book:** Selections from the Prose Writings of ROBERT BUCHANAN.

**The Earthquake;** or, Six Days and a Sabbath. Cr. 8vo, cloth extra, 6s.

**Robert Buchanan's Complete Poetical Works.** With Steel-plate Portrait. Crown 8vo, cloth extra, 7s. 6d.

Crown 8vo, cloth extra, 3s. 6d. each; post 8vo, illust. boards, 2s. each.

**The Shadow of the Sword.**

**A Child of Nature.** With a Frontispiece.

**God and the Man.** With Illustrations by FRED. BARNARD.

**The Martyrdom of Madeline.** With Frontispiece by A. W. COOPER.

**Love Me for Ever.** With a Frontispiece by P. MACNAB.

**Annan Water.**

**The New Abelard.**

**Foxglove Manor.**

**Matt:** A Story of a Caravan.

**The Master of the Mine.** With a Frontispiece by W. H. OVEREND. Crown 8vo, cloth extra, 3s. 6d.

## Bunyan's Pilgrim's Progress.

Edited by Rev. T. SCOTT. With 17 Steel Plates by STOTHARD engraved by GOODALL, and numerous Woodcuts. Crown 8vo, cloth extra, gilt, 7s. 6d.

## Burnett (Mrs.), Novels by:

**Surly Tim,** and other Stories. Post 8vo, illustrated boards, 2s.

Fcap. 8vo, picture cover, 1s. each.

**Kathleen Mavourneen.**

**Lindsay's Luck.**

**Pretty Polly Pemberton.**

**Burton (Captain), Works by:**

**To the Gold Coast for Gold:** A Personal Narrative. By RICHARD F. BURTON and VERNEY LOVETT CAMERON. With Maps and Frontispiece. Two Vols., crown 8vo, cloth extra, 21s.

**The Book of the Sword:** Being a History of the Sword and its Use in all Countries, from the Earliest Times. By RICHARD F. BURTON. With over 400 Illustrations. Square 8vo, cloth extra, 32s.

**Burton (Robert):**

**The Anatomy of Melancholy.** A New Edition, complete, corrected and enriched by Translations of the Classical Extracts. Demy 8vo, cloth extra, 7s. 6d.

**Melancholy Anatomised:** Being an Abridgment, for popular use, of BURTON'S ANATOMY OF MELANCHOLY. Post 8vo, cloth limp, 2s. 6d.

**Byron (Lord):**

**Byron's Childe Harold.** An entirely New Edition of this famous Poem, with over One Hundred new Illusts. by leading Artists. (Uniform with the Illustrated Editions of "The Lady of the Lake" and "Marmion.") Elegantly and appropriately bound, small 4to, 16s.

**Byron's Letters and Journals.** With Notices of his Life. By THOMAS MOORE. A Reprint of the Original Edition, newly revised, with Twelve full-page Plates. Crown 8vo, cloth extra, gilt, 7s. 6d.

**Byron's Don Juan.** Complete in One Vol., post 8vo, cloth limp, 2s.

**Caine.—The Shadow of a Crime:** A Novel. By HALL CAINE. Cr. 8vo, cloth extra, 3s. 6d.; post 8vo, illustrated boards, 2s.

**Cameron (Comdr.), Works by:**

**To the Gold Coast for Gold:** A Personal Narrative. By RICHARD F. BURTON and VERNEY LOVETT CAMERON. With Frontispiece and Maps. Two Vols., crown 8vo, cloth extra, 21s.

**The Cruise of the "Black Prince" Privateer,** Commanded by ROBERT HAWKINS, Master Mariner. By Commander V. LOVETT CAMERON, R.N., C.B., D.C.L. With Frontispiece and Vignette by P. MACNAB. Crown 8vo, cl. ex., 5s. [*Sept.* 15.

**Cameron (Mrs. H. Lovett), Novels by:**

Crown 8vo, cloth extra, 3s. 6d. each post 8vo, illustrated boards, 2s. each.

**Juliet's Guardian. | Deceivers Ever.**

**Carlyle (Thomas):**

**On the Choice of Books.** By THOMAS CARLYLE. With a Life of the Author by R. H. SHEPHERD. New and Revised Edition, post 8vo, cloth extra, Illustrated, 1s. 6d.

**The Correspondence of Thomas Carlyle and Ralph Waldo Emerson,** 1834 to 1872. Edited by CHARLES ELIOT NORTON. With Portraits. Two Vols., crown 8vo, cloth extra, 24s.

**Chapman's (George) Works:** Vol. I. contains the Plays complete, including the doubtful ones. Vol. II., the Poems and Minor Translations, with an Introductory Essay by ALGERNON CHARLES SWINBURNE. Vol. III., the Translations of the Iliad and Odyssey. Three Vols., crown 8vo, cloth extra, 18s.; or separately, 6s. each.

**Chatto & Jackson.—A Treatise on Wood Engraving,** Historical and Practical. By WM. ANDREW CHATTO and JOHN JACKSON. With an Additional Chapter by HENRY G. BOHN; and 450 fine Illustrations. A Reprint of the last Revised Edition. Large 4to, half-bound, 28s.

**Chaucer:**

**Chaucer for Children:** A Golden Key. By Mrs. H. R. HAWEIS. With Eight Coloured Pictures and numerous Woodcuts by the Author. New Ed., small 4to, cloth extra, 6s.

**Chaucer for Schools.** By Mrs. H. R. HAWEIS. Demy 8vo, cloth limp, 2s. 6d.

**City (The) of Dream:** A Poem. Fcap. 8vo, cloth extra, 6s. [*In the press.*

**Clodd.—Myths and Dreams.** By EDWARD CLODD, F.R.A.S., Author of "The Childhood of Religions," &c. Crown 8vo, cloth extra, 5s.

**Cobban.—The Cure of Souls:** A Story. By J. MACLAREN COBBAN. Post 8vo, illustrated boards, 2s.

**Coleman.—Curly:** An Actor's Story. By JOHN COLEMAN. Illustrated by J. C. DOLLMAN. Crown 8vo, 1s. cloth, 1s. 6d.

**Collins (Mortimer), Novels by:**

Crown 8vo, cloth extra, 3s. 6d. each; post 8vo, illustrated boards, 2s. each.

**Sweet Anne Page.**

**Transmigration.**

**From Midnight to Midnight.**

**A Fight with Fortune.** Post 8vo, illustrated boards, 2s.

**Collins (Mortimer & Frances), Novels by:**

Crown 8vo, cloth extra, 3s. 6d. each; post 8vo, illustrated boards, 2s. each.

**Blacksmith and Scholar.**
**The Village Comedy.**
**You Play Me False.**

Post 8vo, illustrated boards, 2s. each.

**Sweet and Twenty.**
**Frances.**

**Collins (Wilkie), Novels by:**

Crown 8vo, cloth extra, Illustrated, 3s. 6d. each; post 8vo, illustrated bds., 2s. each; cloth limp, 2s. 6d. each.

**Antonina.** Illust. by Sir JOHN GILBERT.
**Basil.** Illustrated by Sir JOHN GILBERT and J. MAHONEY.
**Hide and Seek.** Illustrated by Sir JOHN GILBERT and J. MAHONEY.
**The Dead Secret.** Illustrated by Sir JOHN GILBERT.
**Queen of Hearts.** Illustrated by Sir JOHN GILBERT.
**My Miscellanies.** With a Steel-plate Portrait of WILKIE COLLINS.
**The Woman in White.** With Illustrations by Sir JOHN GILBERT and F. A. FRASER.
**The Moonstone.** With Illustrations by G. DU MAURIER and F. A. FRASER.
**Man and Wife.** Illust. by W. SMALL.
**Poor Miss Finch.** Illustrated by G. DU MAURIER and EDWARD HUGHES.
**Miss or Mrs.?** With Illustrations by S. L. FILDES and HENRY WOODS.
**The New Magdalen.** Illustrated by G. DU MAURIER and C. S. REINHARDT.
**The Frozen Deep.** Illustrated by G. DU MAURIER and J. MAHONEY.
**The Law and the Lady.** Illustrated by S. L. FILDES and SYDNEY HALL.
**The Two Destinies.**
**The Haunted Hotel.** Illustrated by ARTHUR HOPKINS.
**The Fallen Leaves.**
**Jezebel's Daughter.**
**The Black Robe.**
**Heart and Science:** A Story of the Present Time.
**"I Say No."**

**The Evil Genius:** A Novel. Three Vols., crown 8vo.

**Collins (C. Allston).—The Bar Sinister:** A Story. By C. ALLSTON COLLINS. Post 8vo, illustrated bds., 2s.

**Colman's Humorous Works:** "Broad Grins," "My Nightgown and Slippers," and other Humorous Works, Prose and Poetical, of GEORGE COLMAN. With Life by G. B. BUCKSTONE, and Frontispiece by HOGARTH. Crown 8vo cloth extra, gilt, 7s. 6d.

**Convalescent Cookery:** A Family Handbook. By CATHERINE RYAN. Crown 8vo, 1s.; cloth, 1s. 6d.

**Conway (Moncure D.), Works by:**

**Demonology and Devil-Lore.** Two Vols., royal 8vo, with 65 Illusts., 28s.
**A Necklace of Stories.** Illustrated by W. J. HENNESSY. Square 8vo, cloth extra, 6s.

**Cook (Dutton), Works by:**

Crown 8vo, cloth extra, 6s. each.

**Hours with the Players.** With Steel Plate Frontispiece.
**Nights at the Play:** A View of the English Stage.

**Leo:** A Novel. Post 8vo, illustrated boards, 2s.
**Paul Foster's Daughter.** Crown 8vo, cloth extra, 3s. 6d.; post 8vo, illustrated boards, 2s.

**Copyright.—A Handbook of English and Foreign Copyright in Literary and Dramatic Works.** By SIDNEY JERROLD, of the Middle Temple, Esq., Barrister-at-Law. Post 8vo, cloth limp, 2s. 6d.

**Cornwall.—Popular Romances of the West of England;** or, The Drolls, Traditions, and Superstitions of Old Cornwall. Collected and Edited by ROBERT HUNT, F.R.S. New and Revised Edition, with Additions, and Two Steel-plate Illustrations by GEORGE CRUIKSHANK. Crown 8vo, cloth extra, 7s. 6d.

**Craddock.—The Prophet of the Great Smoky Mountains.** By CHARLES EGBERT CRADDOCK. Post 8vo, illust. bds., 2s.; cloth limp, 2s. 6d.

**Creasy.—Memoirs of Eminent Etonians:** with Notices of the Early History of Eton College. By Sir EDWARD CREASY, Author of "The Fifteen Decisive Battles of the World." Crown 8vo, cloth extra, gilt, with 13 Portraits, 7s. 6d.

**Cruikshank (George):**

**The Comic Almanack.** Complete in Two SERIES: The FIRST from 1835 to 1843; the SECOND from 1844 to 1853. A Gathering of the BEST HUMOUR of THACKERAY, HOOD, MAYHEW, ALBERT SMITH, A'BECKETT, ROBERT BROUGH, &c. With 2,000 Woodcuts and Steel Engravings by CRUIKSHANK, HINE, LANDELLS, &c. Crown 8vo, cloth gilt, two very thick volumes, 7s. 6d. each.

CRUIKSHANK (GEORGE), *continued.*

**The Life of George Cruikshank.** By BLANCHARD JERROLD, Author of "The Life of Napoleon III.," &c. With 84 Illustrations. New and Cheaper Edition, enlarged, with Additional Plates, and a very carefully compiled Bibliography. Crown 8vo, cloth extra, 7s. 6d.

**Robinson Crusoe.** A beautiful reproduction of Major's Edition, with 37 Woodcuts and Two Steel Plates by GEORGE CRUIKSHANK, choicely printed. Crown 8vo, cloth extra, 7s. 6d.

## Cumming (C. F. Gordon), Works by:

Demy 8vo, cloth extra, 8s. 6d. each.

**In the Hebrides.** With Autotype Facsimile and numerous full-page Illustrations.

**In the Himalayas and on the Indian Plains.** With numerous Illustrations.

**Via Cornwall to Egypt.** With a Photogravure Frontispiece. Demy 8vo, cloth extra, 7s. 6d.

## Cussans.—Handbook of Heraldry;

with Instructions for Tracing Pedigrees and Deciphering Ancient MSS., &c. By JOHN E. CUSSANS. Entirely New and Revised Edition, illustrated with over 400 Woodcuts and Coloured Plates. Crown 8vo, cloth extra, 7s. 6d.

## Cyples.—Hearts of Gold: A Novel.

By WILLIAM CYPLES. Crown 8vo, cloth extra, 3s. 6d.; post 8vo, illustrated boards, 2s.

## Daniel. — Merrie England in the Olden Time.

By GEORGE DANIEL. With Illustrations by ROBT. CRUIKSHANK. Crown 8vo, cloth extra, 3s. 6d.

## Daudet.—The Evangelist; or, Port Salvation.

By ALPHONSE DAUDET. Translated by C. HARRY MELTZER. With Portrait of the Author. Crown 8vo, cloth extra, 3s. 6d.; post 8vo, illust. boards, 2s.

## Davenant. — What shall my Son be?

Hints for Parents on the Choice of a Profession or Trade for their Sons. By FRANCIS DAVENANT, M.A. Post 8vo, cloth limp, 2s. 6d.

## Davies (Dr. N. E.), Works by:

Crown 8vo, 1s. each; cloth limp, 1s. 6d. each.

**One Thousand Medical Maxims.**

**Nursery Hints: A Mother's Guide.**

**Aids to Long Life.** Crown 8vo, 2s.; cloth limp, 2s. 6d.

## Davies' (Sir John) Complete Poetical Works,

including Psalms I. to L. in Verse, and other hitherto Unpublished MSS., for the first time Collected and Edited, with Memorial-Introduction and Notes, by the Rev. A. B. GROSART, D.D. Two Vols., crown 8vo, cloth boards, 12s.

## De Maistre.—A Journey Round My Room.

By XAVIER DE MAISTRE. Translated by HENRY ATTWELL. Post 8vo, cloth limp, 2s. 6d.

## De Mille.—A Castle in Spain:

A Novel. By JAMES DE MILLE. With a Frontispiece. Crown 8vo, cloth extra, 3s. 6d.; post 8vo, illust. bds., 2s.

## Derwent (Leith), Novels by:

Crown 8vo, cloth extra, 3s. 6d. each; post 8vo, illustrated boards, 2s. each.

**Our Lady of Tears.**

**Circe's Lovers.**

## Dickens (Charles), Novels by:

Post 8vo, illustrated boards, 2s. each.

| | |
|---|---|
| **Sketches by Boz.** | **Nicholas Nickleby.** |
| **Pickwick Papers.** | **Oliver Twist.** |

**The Speeches of Charles Dickens 1841–1870.** With a New Bibliography, revised and enlarged. Edited and Prefaced by RICHARD HERNE SHEPHERD. Crown 8vo, cloth extra, 6s.—Also a SMALLER EDITION, in the *Mayfair Library.* Post 8vo, cloth limp, 2s. 6d.

**About England with Dickens.** By ALFRED RIMMER. With 57 Illustrations by C. A. VANDERHOOF, ALFRED RIMMER, and others. Sq. 8vo, cloth extra, 10s. 6d.

## Dictionaries:

**A Dictionary of Miracles: Imitative, Realistic, and Dogmatic.** By the Rev. E. C. BREWER, LL.D. Crown 8vo, cloth extra, 7s. 6d.; hf.-bound, 9s.

**The Reader's Handbook of Allusions, References, Plots, and Stories.** By the Rev. E. C. BREWER, LL.D. Fifth Edition, revised throughout, with a New Appendix, containing a Complete English Bibliography. Crown 8vo, 1,400 pages, cloth extra, 7s. 6d.

**Authors and their Works, with the Dates.** Being the Appendices to "The Reader's Handbook," separately printed. By the Rev. Dr. BREWER. Crown 8vo, cloth limp, 2s.

DICTIONARIES, *continued*—

**Familiar Allusions:** A Handbook of Miscellaneous Information; including the Names of Celebrated Statues, Paintings, Palaces, Country Seats, Ruins, Churches, Ships, Streets, Clubs, Natural Curiosities, and the like. By WM. A. WHEELER and CHARLES G. WHEELER. Demy 8vo, cloth extra, 7s. 6d.

**Short Sayings of Great Men.** With Historical and Explanatory Notes. By SAMUEL A. BENT, M.A. Demy 8vo, cloth extra, 7s. 6d.

**A Dictionary of the Drama:** Being a comprehensive Guide to the Plays, Playwrights, Players, and Playhouses of the United Kingdom and America, from the Earliest to the Present Times. By W. DAVENPORT ADAMS. A thick volume, crown 8vo, half-bound, 12s. 6d. [*In preparation.*

**The Slang Dictionary:** Etymological, Historical, and Anecdotal. Crown 8vo, cloth extra, 6s. 6d.

**Women of the Day:** A Biographical Dictionary. By FRANCES HAYS. Cr. 8vo, cloth extra, 5s.

**Words, Facts, and Phrases:** A Dictionary of Curious, Quaint, and Out-of-the-Way Matters. By ELIEZER EDWARDS. New and Cheaper Issue. Cr. 8vo, cl. ex., 7s. 6d.; hf.-bd., 9s.

**Diderot.—The Paradox of Acting.** Translated, with Annotations, from Diderot's "Le Paradoxe sur le Comédien," by WALTER HERRIES POLLOCK. With a Preface by HENRY IRVING. Cr. 8vo, in parchment, 4s. 6d.

**Dobson (W. T.), Works by:**

Post 8vo, cloth limp, 2s. 6d. each.

**Literary Frivolities, Fancies, Follies, and Frolics.**

**Poetical Ingenuities and Eccentricities.**

**Doran.—Memories of our Great Towns;** with Anecdotic Gleanings concerning their Worthies and their Oddities. By Dr. JOHN DORAN, F.S.A. With 38 Illustrations. New and Cheaper Ed., cr. 8vo, cl. ex., 7s. 6d.

**Drama, A Dictionary of the.** Being a comprehensive Guide to the Plays, Playwrights, Players, and Playhouses of the United Kingdom and America, from the Earliest to the Present Times. By W. DAVENPORT ADAMS. (Uniform with BREWER'S "Reader's Handbook.") Crown 8vo, half-bound, 12s. 6d. [*In preparation.*

**Dramatists, The Old.** Cr. 8vo, cl. ex., Vignette Portraits, 6s. per Vol.

**Ben Jonson's Works.** With Notes Critical and Explanatory, and a Biographical Memoir by WM. GIFFORD. Edit. by Col. CUNNINGHAM. 3 Vols.

**Chapman's Works.** Complete in Three Vols. Vol. I. contains the Plays complete, including doubtful ones; Vol. II., Poems and Minor Translations, with Introductory Essay by A. C. SWINBURNE; Vol. III., Translations of the Iliad and Odyssey.

**Marlowe's Works.** Including his Translations. Edited, with Notes and Introduction, by Col. CUNNINGHAM. One Vol.

**Massinger's Plays.** From the Text of WILLIAM GIFFORD. Edited by Col. CUNNINGHAM. One Vol.

**Dyer.—The Folk-Lore of Plants.** By Rev. T. F. THISELTON DYER, M.A. Crown 8vo, cloth extra, 7s. 6d. [*In preparation.*

**Early English Poets.** Edited, with Introductions and Annotations, by Rev. A. B. GROSART, D.D. Crown 8vo, cloth boards, 6s. per Volume.

**Fletcher's (Giles, B.D.) Complete Poems.** One Vol.

**Davies' (Sir John) Complete Poetical Works.** Two Vols.

**Herrick's (Robert) Complete Collected Poems.** Three Vols.

**Sidney's (Sir Philip) Complete Poetical Works.** Three Vols.

**Herbert (Lord) of Cherbury's Poems.** Edited, with Introduction, by J. CHURTON COLLINS. Crown 8vo, parchment, 8s.

**Edwardes (Mrs. A.), Novels by:**

**A Point of Honour.** Post 8vo, illustrated boards, 2s.

**Archie Lovell.** Crown 8vo, cloth extra, 3s. 6d.; post 8vo, illust. bds., 2s.

**Eggleston.—Roxy:** A Novel. By EDWARD EGGLESTON. Post 8vo, illust. boards, 2s.

**Emanuel.—On Diamonds and Precious Stones:** their History, Value, and Properties; with Simple Tests for ascertaining their Reality. By HARRY EMANUEL, F.R.G.S. With numerous Illustrations, tinted and plain. Crown 8vo, cloth extra, gilt, 6s.

**Englishman's House, The:** A Practical Guide to all interested in Selecting or Building a House, with full Estimates of Cost, Quantities, &c. By C. J. RICHARDSON. Third Edition. Nearly 600 Illusts. Cr. 8vo, cl. ex., 7s. 6d.

**English Merchants:** Memoirs in Illustration of the Progress of British Commerce. By H. R. FOX BOURNE. With Illusts. New and Cheaper Edit. revised. Crown 8vo, cloth extra, 7s. 6d.

**Ewald (Alex. Charles, F.S.A.), Works by:**

**The Life and Times of Prince Charles Stuart,** Count of Albany, commonly called the Young Pretender. From the State Papers and other Sources. New and Cheaper Edition, with a Portrait, crown 8vo, cloth extra, 7s. 6d.

**Stories from the State Papers.** With an Autotype Facsimile. Crown 8vo, cloth extra, 6s.

**Studies Re-studied:** Historical Sketches from Original Sources. Demy 8vo cloth extra, 12s.

**Eyes, The.—How to Use our** Eyes, and How to Preserve Them. By JOHN BROWNING, F.R.A.S., &c. Fourth Edition. With 55 Illustrations. Crown 8vo, cloth, 1s.

**Fairholt.—Tobacco:** Its History and Associations; with an Account of the Plant and its Manufacture, and its Modes of Use in all Ages and Countries. By F. W. FAIRHOLT, F.S.A. With upwards of 100 Illustrations by the Author. Crown 8vo, cloth extra, 6s.

**Familiar Allusions:** A Handbook of Miscellaneous Information; including the Names of Celebrated Statues, Paintings, Palaces, Country Seats, Ruins, Churches, Ships, Streets, Clubs, Natural Curiosities, and the like. By WILLIAM A. WHEELER, Author of "Noted Names of Fiction;" and CHARLES G. WHEELER. Demy 8vo, cloth extra, 7s. 6d.

**Faraday (Michael), Works by:**

Post 8vo, cloth extra, 4s. 6d. each.

**The Chemical History of a Candle:** Lectures delivered before a Juvenile Audience at the Royal Institution. Edited by WILLIAM CROOKES, F.C.S. With numerous Illustrations.

**On the Various Forces of Nature,** and their Relations to each other: Lectures delivered before a Juvenile Audience at the Royal Institution. Edited by WILLIAM CROOKES, F.C.S. With numerous Illustrations.

**Farrer. — Military Manners and Customs.** By J. A. FARRER, Author of "Primitive Manners and Customs," &c. Cr. 8vo, cloth extra, 6s.

**Fin-Bec. — The Cupboard Papers:** Observations on the Art of Living and Dining. By FIN-BEC. Post 8vo, cloth limp, 2s. 6d.

**Fitzgerald (Percy), Works by:**

**The Recreations of a Literary Man;** or, Does Writing Pay? With Recollections of some Literary Men, and a View of a Literary Man's Working Life. Cr. 8vo, cloth extra, 6s.

**The World Behind the Scenes.** Crown 8vo, cloth extra, 3s. 6d.

**Little Essays:** Passages from the Letters of CHARLES LAMB. Post 8vo, cloth limp, 2s. 6d.

Post 8vo, illustrated boards, 2s. each.

**Bella Donna.** | **Never Forgotten**

**The Second Mrs. Tillotson.**

**Polly.**

**Seventy-five Brooke Street.**

**The Lady of Brantome.**

**Fletcher's (Giles, B.D.) Complete Poems:** Christ's Victorie in Heaven, Christ's Victorie on Earth, Christ's Triumph over Death, and Minor Poems. With Memorial-Introduction and Notes by the Rev. A. B. GROSART, D.D. Cr. 8vo, cloth bds., 6s.

**Fonblanque.—Filthy Lucre:** A Novel. By ALBANY DE FONBLANQUE. Post 8vo, illustrated boards, 2s.

**Francillon (R. E.), Novels by:**

Crown 8vo, cloth extra, 3s. 6d. each; post 8vo, illust. boards, 2s. each.

**One by One.** | **A Real Queen.**

**Queen Cophetua.**

**Olympia.** Post 8vo, illust. boards, 2s.

**Esther's Glove.** Fcap. 8vo, 1s.

**French Literature, History of.** By HENRY VAN LAUN. Complete in 3 Vols., demy 8vo, cl. bds., 7s. 6d. each.

**Frere.—Pandurang Hari;** or, Memoirs of a Hindoo. With a Preface by Sir H. BARTLE FRERE, G.C.S.I., &c. Crown 8vo, cloth extra, 3s. 6d.; post 8vo, illustrated boards, 2s.

**Friswell.—One of Two:** A Novel. By HAIN FRISWELL. Post 8vo, illustrated boards, 2s.

**Frost (Thomas), Works by:**

Crown 8vo, cloth extra, 3s. 6d. each.

**Circus Life and Circus Celebrities.**

**The Lives of the Conjurers.**

**The Old Showmen and the Old London Fairs.**

**Fry's (Herbert) Royal Guide to the London Charities, 1886-7.** Showing their Name, Date of Foundation, Objects, Income, Officials, &c. Published Annually. Cr. 8vo, cloth, 1s. 6d.

**Gardening Books:**
Post 8vo, 1s. each; cl. limp, 1s. 6d. each.
**A Year's Work in Garden and Greenhouse:** Practical Advice to Amateur Gardeners as to the Management of the Flower, Fruit, and Frame Garden. By GEORGE GLENNY.
**Our Kitchen Garden:** The Plants we Grow, and How we Cook Them. By TOM JERROLD.
**Household Horticulture:** A Gossip about Flowers. By TOM and JANE JERROLD. Illustrated.
**The Garden that Paid the Rent.** By TOM JERROLD.

**My Garden Wild, and What I Grew there.** By F. G. HEATH. Crown 8vo, cloth extra, 5s.; gilt edges, 6s.

**Garrett.—The Capel Girls:** A Novel. By EDWARD GARRETT. Cr. 8vo, cl. ex., 3s. 6d.; post 8vo, illust. bds., 2s.

**Gentleman's Magazine (The) for 1886.** One Shilling Monthly. In addition to the Articles upon subjects in Literature, Science, and Art, for which this Magazine has so high a reputation, "**Science Notes,**" by W. MATTIEU WILLIAMS, F.R.A.S., and "**Table Talk,**" by SYLVANUS URBAN, appear monthly.
*** *Now ready, the Volume for* JANUARY *to* JUNE, 1886, *cloth extra, price* 8s. 6d.; *Cases for binding,* 2s. *each.*

**Gentleman's Annual (The) for Christmas, 1886.** Containing a Complete Novel, "**Wife or No Wife?**" by T. W. SPEIGHT, Author of "The Mysteries of Heron Dyke." Demy 8vo, 1s. [*Preparing.*

**German Popular Stories.** Collected by the Brothers GRIMM, and Translated by EDGAR TAYLOR. Edited, with an Introduction, by JOHN RUSKIN. With 22 Illustrations on Steel by GEORGE CRUIKSHANK. Square 8vo, cloth extra, 6s. 6d.; gilt edges, 7s. 6d.

**Gibbon (Charles), Novels by:**
Crown 8vo, cloth extra, 3s. 6d. each post 8vo, illustrated boards, 2s. each.

**Robin Gray.**
**For Lack of Gold.**
**What will the World Say?**
**In Honour Bound.**
**Queen of the Meadow.**
**Braes of Yarrow.**
**The Flower of the Forest.**
**A Heart's Problem.**
**The Golden Shaft.**
**Of High Degree.**
**Fancy Free.**

Post 8vo, illustrated boards, 2s. each.
**For the King.** | **In Pastures Green**
**In Love and War.**
**By Mead and Stream.**
**Heart's Delight.** [*Preparing.*

Crown 8vo, cloth extra, 3s. 6d. each.
**Loving a Dream.** | **A Hard Knot.**

**Gilbert (William), Novels by:**
Post 8vo, illustrated boards, 2s. each.
**Dr. Austin's Guests.**
**The Wizard of the Mountain.**
**James Duke, Costermonger.**

**Gilbert (W. S.), Original Plays by:** In Two Series, each complete in itself, price 2s. 6d. each.
The FIRST SERIES contains—The Wicked World—Pygmalion and Galatea—Charity—The Princess—The Palace of Truth—Trial by Jury.
The SECOND SERIES contains—Broken Hearts—Engaged—Sweethearts—Gretchen—Dan'l Druce—Tom Cobb—H.M.S. Pinafore—The Sorcerer—The Pirates of Penzance.

**Eight Original Comic Operas.** Written by W. S. GILBERT. Containing: The Sorcerer—H.M.S. "Pinafore"—The Pirates of Penzance—Iolanthe—Patience—Princess Ida—The Mikado—Trial by Jury. Demy 8vo, cloth limp, 2s. 6d.

**Glenny.—A Year's Work in Garden and Greenhouse:** Practical Advice to Amateur Gardeners as to the Management of the Flower, Fruit, and Frame Garden. By GEORGE GLENNY. Post 8vo, 1s.; cloth, 1s. 6d.

**Godwin.—Lives of the Necromancers.** By WILLIAM GODWIN. Post 8vo, cloth limp, 2s.

**Golden Library, The:**
Square 16mo (Tauchnitz size), cloth limp, 2s. per volume.
**Bayard Taylor's Diversions of the Echo Club.**
**Bennett's (Dr. W. C.) Ballad History of England.**
**Bennett's (Dr.) Songs for Sailors.**
**Byron's Don Juan.**
**Godwin's (William) Lives of the Necromancers.**
**Holmes's Autocrat of the Breakfast Table.** Introduction by SALA.
**Holmes's Professor at the Breakfast Table.**
**Hood's Whims and Oddities.** Complete. All the original Illustrations.
**Irving's (Washington) Tales of a Traveller.**
**Jesse's (Edward) Scenes and Occupations of a Country Life.**
**Lamb's Essays of Elia.** Both Series Complete in One Vol.
**Leigh Hunt's Essays:** A Tale for a Chimney Corner, and other Pieces. With Portrait, and Introduction by EDMUND OLLIER.

GOLDEN LIBRARY, *continued.*

Mallory's (Sir Thomas) Mort d'Arthur: The Stories of King Arthur and of the Knights of the Round Table. Edited by B. MONTGOMERIE RANKING.

Pascal's Provincial Letters. A New Translation, with Historical Introduction and Notes, by T. M'CRIE, D.D.

Pope's Poetical Works. Complete.

Rochefoucauld's Maxims and Moral Reflections. With Notes, and Introductory Essay by SAINTE-BEUVE.

St. Pierre's Paul and Virginia, and The Indian Cottage. Edited, with Life, by the Rev. E. CLARKE.

Shelley's Early Poems, and Queen Mab. With Essay by LEIGH HUNT.

Shelley's Later Poems: Laon and Cythna, &c.

Shelley's Posthumous Poems, the Shelley Papers, &c.

Shelley's Prose Works, including A Refutation of Deism, Zastrozzi, St. Irvyne, &c.

**Golden Treasury of Thought, The:** An ENCYCLOPÆDIA OF QUOTATIONS from Writers of all Times and Countries. Selected and Edited by THEODORE TAYLOR. Crown 8vo, cloth gilt and gilt edges, 7s. 6d.

**Graham. — The Professor's Wife:** A Story. By LEONARD GRAHAM. Fcap. 8vo, picture cover, 1s.

**Greeks and Romans, The Life of the,** Described from Antique Monuments. By ERNST GUHL and W. KONER. Translated from the Third German Edition, and Edited by Dr. F. HUEFFER. 545 Illusts. New and Cheaper Edit., demy 8vo, cl. ex., 7s. 6d.

**Greenaway (Kate) and Bret Harte.—The Queen of the Pirate Isle.** By BRET HARTE. With 25 original Drawings by KATE GREENAWAY, Reproduced in Colours by E. EVANS. Sm. 4to, bds., 5s. [*Shortly.*

**Greenwood (James), Works by:**

Crown 8vo, cloth extra, 3s. 6d. each.

The Wilds of London.

Low-Life Deeps: An Account of the Strange Fish to be Found There.

Dick Temple: A Novel. Post 8vo, illustrated boards, 2s.

**Guyot.—The Earth and Man;** or, Physical Geography in its relation to the History of Mankind. By ARNOLD GUYOT. With Additions by Professors AGASSIZ, PIERCE, and GRAY; 12 Maps and Engravings on Steel, some Coloured, and copious Index. Crown 8vo, cloth extra, gilt, 4s. 6d.

**Hair (The): Its Treatment in** Health, Weakness, and Disease. Translated from the German of Dr. J. PINCUS. Crown 8vo, 1s.; cloth, 1s. 6d.

**Hake (Dr. Thomas Gordon),** Poems by:

Crown 8vo, cloth extra, 6s. each.

New Symbols.

Legends of the Morrow.

The Serpent Play.

Maiden Ecstasy. Small 4to, cloth extra, 8s.

**Hall.—Sketches of Irish Character.** By Mrs. S. C. HALL. With numerous Illustrations on Steel and Wood by MACLISE, GILBERT, HARVEY, and G. CRUIKSHANK. Medium 8vo, cloth extra, gilt, 7s. 6d.

**Halliday.—Every-day Papers.** By ANDREW HALLIDAY. Post 8vo, illustrated boards, 2s.

**Handwriting, The Philosophy of.** With over 100 Facsimiles and Explanatory Text. By DON FELIX DE SALAMANCA. Post 8vo, cl. limp, 2s. 6d.

**Hanky-Panky: A Collection of** Very Easy Tricks, Very Difficult Tricks, White Magic, Sleight of Hand, &c. Edited by W. H. CREMER. With 200 Illusts. Crown 8vo, cloth extra, 4s. 6d.

**Hardy (Lady Duffus). — Paul Wynter's Sacrifice:** A Story. By Lady DUFFUS HARDY. Post 8vo, illust. boards, 2s.

**Hardy (Thomas).—Under the Greenwood Tree.** By THOMAS HARDY, Author of "Far from the Madding Crowd." With numerous Illustrations. Crown 8vo, cloth extra, 3s. 6d.; post 8vo, illustrated boards, 2s.

**Harwood.—The Tenth Earl.** By J. BERWICK HARWOOD. Post 8vo, illustrated boards, 2s.

**Haweis (Mrs. H. R.), Works by:**

The Art of Dress. With numerous Illustrations. Small 8vo, illustrated cover, 1s.; cloth limp, 1s. 6d.

The Art of Beauty. New and Cheaper Edition. Crown 8vo, cloth extra, Coloured Frontispiece and Illusts. 6s.

The Art of Decoration. Square 8vo, handsomely bound and profusely Illustrated, 10s. 6d.

Chaucer for Children: A Golden Key. With Eight Coloured Pictures and numerous Woodcuts. New Edition, small 4to, cloth extra, 6s.

Chaucer for Schools. Demy 8vo, cloth limp, 2s. 6d.

**Haweis (Rev. H. R.).—American Humorists.** Including WASHINGTON IRVING, OLIVER WENDELL HOLMES, JAMES RUSSELL LOWELL, ARTEMUS WARD, MARK TWAIN, and BRET HARTE. By the Rev. H. R. HAWEIS, M.A. Crown 8vo, cloth extra, 6s.

**Hawthorne (Julian), Novels by.** Crown 8vo, cloth extra, 3s. 6d. each; post 8vo, illustrated boards, 2s. each.

**Garth.** | **Sebastian Strome.**
**Ellice Quentin.** | **Dust.**
**Prince Saroni's Wife.**
**Fortune's Fool.** | **Beatrix Randolph.**

Crown 8vo, cloth extra, 3s. 6d. each.
**Miss Cadogna.**
**Love—or a Name.**

**Mrs. Gainsborough's Diamonds.** Fcap. 8vo, illustrated cover, 1s.

**Hays.—Women of the Day:** A Biographical Dictionary of Notable Contemporaries. By FRANCES HAYS. Crown 8vo, cloth extra, 5s.

**Heath (F. G.). — My Garden Wild,** and What I Grew There. By FRANCIS GEORGE HEATH, Author of "The Fern World," &c. Crown 8vo, cloth extra, 5s.; cl. gilt, gilt edges, 6s.

**Helps (Sir Arthur), Works by:**
Post 8vo, cloth limp, 2s. 6d. each.
**Animals and their Masters.**
**Social Pressure.**

**Ivan de Biron:** A Novel. Crown 8vo, cloth extra, 3s. 6d.; post 8vo, illustrated boards, 2s.

**Heptalogia (The); or, The** Seven against Sense. A Cap with Seven Bells. Cr. 8vo, cloth extra, 6s.

**Herrick's (Robert) Hesperides,** Noble Numbers, and Complete Collected Poems. With Memorial-Introduction and Notes by the Rev. A. B. GROSART, D.D., Steel Portrait, Index of First Lines, and Glossarial Index, &c. Three Vols., crown 8vo, cloth, 18s.

**Hesse-Wartegg (Chevalier Ernst von), Works by:**

**Tunis:** The Land and the People. With 22 Illustrations. Crown 8vo, cloth extra, 3s. 6d.

**The New South-West:** Travelling Sketches from Kansas, New Mexico, Arizona, and Northern Mexico. With 100 fine Illustrations and Three Maps. Demy 8vo, cloth extra, 14s. [*In preparation.*

**Herbert.—The Poems of Lord Herbert of Cherbury.** Edited, with Introduction, by J. CHURTON COLLINS. Crown 8vo, bound in parchment, 8s.

**Hindley (Charles), Works by:**
Crown 8vo, cloth extra, 3s. 6d. each.

**Tavern Anecdotes and Sayings:** Including the Origin of Signs, and Reminiscences connected with Taverns, Coffee Houses, Clubs, &c. With Illustrations.

**The Life and Adventures of a Cheap Jack.** By One of the Fraternity. Edited by CHARLES HINDLEY.

**Hoey.—The Lover's Creed.** By Mrs. CASHEL HOEY. With Frontispiece by P. MACNAB. New and Cheaper Edit. Crown 8vo, cloth extra, 3s. 6d.; post 8vo, illustrated boards, 2s.

**Holmes (O. Wendell), Works by:**

**The Autocrat of the Breakfast-Table.** Illustrated by J. GORDON THOMSON. Post 8vo, cloth limp, 2s. 6d.—Another Edition in smaller type, with an Introduction by G. A. SALA. Post 8vo, cloth limp, 2s.

**The Professor at the Breakfast-Table;** with the Story of Iris. Post 8vo, cloth limp, 2s.

**Holmes. — The Science of Voice Production and Voice Preservation:** A Popular Manual for the Use of Speakers and Singers. By GORDON HOLMES, M.D. With Illustrations. Crown 8vo, 1s.; cloth, 1s. 6d.

**Hood (Thomas):**

**Hood's Choice Works,** in Prose and Verse. Including the Cream of the COMIC ANNUALS. With Life of the Author, Portrait, and 200 Illustrations. Crown 8vo, cloth extra, 7s. 6d.

**Hood's Whims and Oddities.** Complete. With all the original Illustrations. Post 8vo, cloth limp, 2s.

**Hood (Tom), Works by:**

**From Nowhere to the North Pole:** A Noah's Arkæological Narrative. With 25 Illustrations by W. BRUNTON and E. C. BARNES. Square crown 8vo, cloth extra, gilt edges, 6s.

**A Golden Heart:** A Novel. Post 8vo, illustrated boards, 2s.

**Hook's (Theodore) Choice Humorous Works,** including his Ludicrous Adventures, Bons Mots, Puns and Hoaxes. With a New Life of the Author, Portraits, Facsimiles, and Illusts. Cr. 8vo, cl. extra, gilt, 7s. 6d.

**Hooper.—The House of Raby:** A Novel. By Mrs. GEORGE HOOPER. Post 8vo, illustrated boards, 2s.

**Hopkins—"'Twixt Love and Duty:"** A Novel. By TIGHE HOPKINS. Crown 8vo, cloth extra, 6s.

**Horne.—Orion:** An Epic Poem, in Three Books. By RICHARD HENGIST HORNE. With Photographic Portrait from a Medallion by SUMMERS. Tenth Edition, crown 8vo, cloth extra, 7s.

**Howell.—Conflicts of Capital and Labour,** Historically and Economically considered: Being a History and Review of the Trade Unions of Great Britain. By GEO. HOWELL M.P. Crown 8vo, cloth extra, 7s. 6d.

**Hugo. — The Hunchback of Notre Dame.** By VICTOR HUGO. Post 8vo, illustrated boards, 2s.

**Hunt.—Essays by Leigh Hunt.** A Tale for a Chimney Corner, and other Pieces. With Portrait and Introduction by EDMUND OLLIER. Post 8vo, cloth limp, 2s.

**Hunt (Mrs. Alfred), Novels by:**

Crown 8vo, cloth extra, 3s. 6d. each post 8vo, illustrated boards, 2s. each.

**Thornicroft's Model.**

**The Leaden Casket.**

**Self-Condemned**

**That other Person.** Three Vols., crown 8vo. [*Shortly.*

**Indoor Paupers.** By ONE OF THEM. Crown 8vo, 1s.; cloth, 1s. 6d.

**Ingelow.—Fated to be Free:** A Novel. By JEAN INGELOW. Crown 8vo, cloth extra, 3s. 6d.; post 8vo, illustrated boards, 2s.

**Irish Wit and Humour, Songs of.** Collected and Edited by A. PERCEVAL GRAVES. Post 8vo, cloth limp, 2s. 6d.

**Irving—Tales of a Traveller.** By WASHINGTON IRVING. Post 8vo, cloth limp, 2s.

**Jay (Harriett), Novels by:**

**The Dark Colleen.** Post 8vo, illustrated boards, 2s.

**The Queen of Connaught.** Crown 8vo, cloth extra, 3s. 6d.; post 8vo, illustrated boards, 2s.

**Janvier.—Practical Keramics for Students.** By CATHERINE A. JANVIER. Crown 8vo, cloth extra, 6s.

**Jefferies (Richard), Works by:**

Crown 8vo, cloth extra, 6s. each.

**Nature near London.**

**The Life of the Fields.**

**The Open Air.**

**Jennings (Hargrave). — The Rosicrucians:** Their Rites and Mysteries. With Chapters on the Ancient Fire and Serpent Worshippers. By HARGRAVE JENNINGS. With Five full-page Plates and upwards of 300 Illustrations. A New Edition, crown 8vo, cloth extra, 7s. 6d.

**Jennings (H. J.), Works by:**

**Curiosities of Criticism.** Post 8vo, cloth limp, 2s. 6d.

**Lord Tennyson:** A Biographical Sketch. With a Photograph-Portrait. Crown 8vo, cloth extra, 6s.

**Jerrold (Tom), Works by:**

Post 8vo, 1s. each; cloth, 1s. 6d. each.

**The Garden that Paid the Rent.**

**Household Horticulture:** A Gossip about Flowers. Illustrated.

**Our Kitchen Garden:** The Plants we Grow, and How we Cook Them.

**Jesse.—Scenes and Occupations of a Country Life.** By EDWARD JESSE. Post 8vo, cloth limp, 2s.

**Jeux d'Esprit.** Collected and Edited by HENRY S. LEIGH. Post 8vo, cloth limp, 2s. 6d.

**Jones (Wm., F.S.A.), Works by:**

Crown 8vo, cloth extra, 7s. 6d. each.

**Finger-Ring Lore:** Historical, Legendary, and Anecdotal. With over Two Hundred Illustrations.

**Credulities, Past and Present;** including the Sea and Seamen, Miners, Talismans, Word and Letter Divination, Exorcising and Blessing of Animals, Birds, Eggs, Luck, &c. With an Etched Frontispiece.

**Crowns and Coronations:** A History of Regalia in all Times and Countries. With One Hundred Illustrations.

**Jonson's (Ben) Works.** With Notes Critical and Explanatory, and a Biographical Memoir by WILLIAM GIFFORD. Edited by Colonel CUNNINGHAM. Three Vols., crown 8vo, cloth extra, 18s.; or separately, 6s. each.

**Maclise Portrait-Gallery (The) of Illustrious Literary Characters;** with Memoirs—Biographical, Critical, Bibliographical, and Anecdotal—illustrative of the Literature of the former half of the Present Century. By WILLIAM BATES, B.A. With 85 Portraits printed on an India Tint. Crown 8vo, cloth extra, 7s. 6d.

**Mackay.—Interludes and Undertones:** or, Music at Twilight. By CHARLES MACKAY, LL.D. Crown 8vo, cloth extra, 6s.

**Macquoid (Mrs.), Works by:**
Square 8vo, cloth extra, 10s. 6d. each.
**In the Ardennes.** With 50 fine Illustrations by THOMAS R. MACQUOID.
**Pictures and Legends from Normandy and Brittany.** With numerous Illustrations by THOMAS R. MACQUOID.
**About Yorkshire.** With 67 Illustrations by T. R. MACQUOID.

Crown 8vo, cloth extra, 7s. 6d each.
**Through Normandy.** With 90 Illustrations by T. R. MACQUOID.
**Through Brittany.** With numerous Illustrations by T. R. MACQUOID.

Post 8vo, illustrated boards, 2s. each.
**The Evil Eye, and other Stories.**
**Lost Rose.**

**Magician's Own Book (The):** Performances with Cups and Balls, Eggs, Hats, Handkerchiefs, &c. All from actual Experience. Edited by W. H. CREMER. With 200 Illustrations. Crown 8vo, cloth extra, 4s. 6d.

**Magic Lantern (The), and its Management:** including full Practical Directions for producing the Limelight, making Oxygen Gas, and preparing Lantern Slides. By T. C. HEPWORTH. With 10 Illustrations. Crown 8vo, 1s.; cloth, 1s. 6d.

**Magna Charta.** An exact Facsimile of the Original in the British Museum, printed on fine plate paper, 3 feet by 2 feet, with Arms and Seals emblazoned in Gold and Colours. 5s.

**Mallock (W. H.), Works by:**
**The New Republic;** or, Culture, Faith and Philosophy in an English Country House. Post 8vo, cloth limp, 2s. 6d.; Cheap Edition, illustrated boards, 2s.
**The New Paul and Virginia;** or, Positivism on an Island. Post 8vo, cloth limp, 2s. 6d.
**Poems.** Small 4to, in parchment, 8s.
**Is Life worth Living?** Crown 8vo, cloth extra, 6s.

**Mallory's (Sir Thomas) Mort d'Arthur:** The Stories of King Arthur and of the Knights of the Round Table. Edited by B. MONTGOMERIE RANKING. Post 8vo, cloth limp, 2s.

**Marlowe's Works.** Including his Translations. Edited, with Notes and Introductions, by Col. CUNNINGHAM. Crown 8vo, cloth extra, 6s.

**Marryat (Florence), Novels by:**
Crown 8vo, cloth extra, 3s. 6d. each; post 8vo, illustrated boards, 2s. each.
**Open! Sesame!**
**Written in Fire**

Post 8vo, illustrated boards, 2s. each.
**A Harvest of Wild Oats.**
**A Little Stepson.**
**Fighting the Air.**

**Masterman.—Half a Dozen Daughters:** A Novel. By J. MASTERMAN. Post 8vo, illustrated boards, 2s.

**Mark Twain, Works by:**
**The Choice Works of Mark Twain.** Revised and Corrected throughout by the Author. With Life, Portrait, and numerous Illustrations. Crown 8vo, cloth extra, 7s. 6d.
**The Innocents Abroad;** or, The New Pilgrim's Progress: Being some Account of the Steamship "Quaker City's" Pleasure Excursion to Europe and the Holy Land. With 234 Illustrations. Crown 8vo, cloth extra, 7s. 6d.—Cheap Edition (under the title of "MARK TWAIN'S PLEASURE TRIP"), post 8vo, illust. boards, 2s.
**Roughing It, and The Innocents at Home.** With 200 Illustrations by F. A. FRASER. Crown 8vo, cloth extra, 7s. 6d.
**The Gilded Age.** By MARK TWAIN and CHARLES DUDLEY WARNER. With 212 Illustrations by T. COPPIN. Crown 8vo, cloth extra, 7s. 6d.
**The Adventures of Tom Sawyer.** With 111 Illustrations. Crown 8vo, cloth extra, 7s. 6d.—Cheap Edition, post 8vo, illustrated boards, 2s.
**The Prince and the Pauper.** With nearly 200 Illustrations. Crown 8vo, cloth extra, 7s. 6d.
**A Tramp Abroad.** With 314 Illustrations. Crown 8vo, cloth extra, 7s. 6d.—Cheap Edition, post 8vo, illustrated boards, 2s.
**The Stolen White Elephant, &c.** Crown 8vo, cloth extra, 6s.; post 8vo, illustrated boards, 2s.

MARK TWAIN'S WORKS, *continued*—

**Life on the Mississippi.** With about 300 Original Illustrations. Crown 8vo, cloth extra, 7s. 6d.

**The Adventures of Huckleberry Finn.** With 174 Illustrations by E. W. KEMBLE. Crown 8vo, cloth extra, 7s. 6d.—Cheap Edition, post 8vo, illustrated boards, 2s.

## Massinger's Plays.
From the Text of WILLIAM GIFFORD. Edited by Col. CUNNINGHAM. Crown 8vo, cloth extra, 6s.

## Matthews.—A Secret of the Sea, &c.
By BRANDER MATTHEWS. Post 8vo, illustrated boards, 2s.; cloth, 2s. 6d.

## Mayfair Library, The:
Post 8vo, cloth limp, 2s. 6d. per Volume.

**A Journey Round My Room.** By XAVIER DE MAISTRE. Translated by HENRY ATTWELL.

**Latter-Day Lyrics.** Edited by W DAVENPORT ADAMS.

**Quips and Quiddities.** Selected by W. DAVENPORT ADAMS.

**The Agony Column of "The Times,"** from 1800 to 1870. Edited, with an Introduction, by ALICE CLAY.

**Melancholy Anatomised:** A Popular Abridgment of "Burton's Anatomy of Melancholy."

**Gastronomy as a Fine Art.** By BRILLAT-SAVARIN.

**The Speeches of Charles Dickens.**

**Literary Frivolities, Fancies, Follies, and Frolics.** By W. T. DOBSON.

**Poetical Ingenuities and Eccentricities.** Selected and Edited by W. T. DOBSON.

**The Cupboard Papers.** By FIN-BEC.

**Original Plays by W. S. GILBERT.** FIRST SERIES. Containing: The Wicked World — Pygmalion and Galatea—Charity — The Princess—The Palace of Truth—Trial by Jury.

**Original Plays by W. S. GILBERT.** SECOND SERIES. Containing: Broken Hearts — Engaged — Sweethearts—Gretchen—Dan'l Druce—Tom Cobb —H.M.S. Pinafore — The Sorcerer —The Pirates of Penzance.

**Songs of Irish Wit and Humour.** Collected and Edited by A. PERCEVAL GRAVES.

**Animals and their Masters.** By Sir ARTHUR HELPS.

**Social Pressure.** By Sir A. HELPS.

MAYFAIR LIBRARY, *continued*—

**Curiosities of Criticism.** By HENRY J. JENNINGS.

**The Autocrat of the Breakfast-Table.** By OLIVER WENDELL HOLMES. Illustrated by J. GORDON THOMSON.

**Pencil and Palette.** By ROBERT KEMPT.

**Little Essays:** Sketches and Characters. By CHAS. LAMB. Selected from his Letters by PERCY FITZGERALD.

**Forensic Anecdotes;** or, Humour and Curiosities of the Law and Men of Law. By JACOB LARWOOD.

**Theatrical Anecdotes.** By JACOB LARWOOD.

**Jeux d'Esprit.** Edited by HENRY S. LEIGH.

**True History of Joshua Davidson** By E. LYNN LINTON.

**Witch Stories.** By E. LYNN LINTON.

**Ourselves:** Essays on Women. By E. LYNN LINTON.

**Pastimes and Players.** By ROBERT MACGREGOR.

**The New Paul and Virginia** By W. H. MALLOCK.

**New Republic.** By W. H. MALLOCK.

**Puck on Pegasus.** By H. CHOLMONDELEY-PENNELL.

**Pegasus Re-Saddled.** By H. CHOLMONDELEY-PENNELL. Illustrated by GEORGE DU MAURIER.

**Muses of Mayfair.** Edited by H. CHOLMONDELEY-PENNELL.

**Thoreau:** His Life and Aims. By H. A. PAGE.

**Punlana.** By the Hon. HUGH ROWLEY.

**More Punlana.** By the Hon. HUGH ROWLEY.

**The Philosophy of Handwriting.** By DON FELIX DE SALAMANCA.

**By Stream and Sea.** By WILLIAM SENIOR.

**Old Stories Re-told.** By WALTER THORNBURY.

**Leaves from a Naturalist's Note-Book.** By Dr. ANDREW WILSON.

## Mayhew.—London Characters
and the Humorous Side of London Life. By HENRY MAYHEW. With numerous Illustrations. Crown 8vo, cloth extra, 3s. 6d.

## Medicine, Family.—One Thousand Medical Maxims and Surgical Hints,
for Infancy, Adult Life, Middle Age, and Old Age. By N. E. DAVIES, L.R.C.P. Lond. Cr. 8vo, 1s.; cl., 1s. 6d.

**Merry Circle (The):** A Book of New Intellectual Games and Amusements. By CLARA BELLEW. With numerous Illustrations. Crown 8vo, cloth extra, 4s. 6d.

**Mexican Mustang (On a),** through Texas, from the Gulf to the Rio Grande. A New Book of American Humour. By ALEX. E. SWEET and J. ARMOY KNOX, Editors of "Texas Siftings." With 265 Illusts. Cr. 8vo, cloth extra, 7s. 6d.

**Middlemass (Jean), Novels by:** Post 8vo, illustrated boards, 2s. each.
- **Touch and Go.**
- **Mr. Dorillion.**

**Miller.—Physiology for the Young;** or, The House of Life: Human Physiology, with its application to the Preservation of Health. For Classes and Popular Reading. With numerous Illusts. By Mrs. F. FENWICK MILLER. Small 8vo, cloth limp, 2s. 6d

**Milton (J. L.), Works by:**
Sm. 8vo, 1s. each; cloth ex., 1s. 6d. each.
- **The Hygiene of the Skin.** A Concise Set of Rules for the Management of the Skin; with Directions for Diet, Wines, Soaps, Baths, &c.
- **The Bath in Diseases of the Skin.**
- **The Laws of Life,** and their Relation to Diseases of the Skin.

**Molesworth (Mrs.).—Hathercourt Rectory.** By Mrs. MOLESWORTH, Author of "The Cuckoo Clock," &c. Crown 8vo, cloth extra, 4s. 6d.

**Murray (D. Christie), Novels by.** Crown 8vo, cloth extra, 3s. 6d. each; post 8vo, illustrated boards, 2s. each.
- **A Life's Atonement.**
- **A Model Father.**
- **Joseph's Coat.**
- **Coals of Fire.**
- **By the Gate of the Sea.**
- **Val Strange.**
- **Hearts.**
- **The Way of the World.**
- **A Bit of Human Nature.**

Crown 8vo, cloth extra, 3s. 6d. each.
- **First Person Singular:** A Novel. With a Frontispiece by ARTHUR HOPKINS.
- **Cynic Fortune:** A Tale of a Man with a Conscience. With a Frontispiece by R. CATON WOODVILLE.

**North Italian Folk.** By Mrs. COMYNS CARR. Illustrated by RANDOLPH CALDECOTT. Square 8vo, cloth extra, 7s. 6d.

**Number Nip (Stories about),** the Spirit of the Giant Mountains. Retold for Children by WALTER GRAHAME. With Illustrations by J. MOYR SMITH. Post 8vo, cl. extra, 5s.

**Nursery Hints:** A Mother's Guide in Health and Disease. By N. E. DAVIES, L.R.C.P. Crown 8vo, 1s cloth, 1s. 6d.

**O'Connor.—Lord Beaconsfield** A Biography. By T. P. O'CONNOR, M.P. Sixth Edition, with a New Preface, bringing the work down to the Death of Lord Beaconsfield. Crown 8vo, cloth extra, 7s. 6d.

**O'Hanlon.—The Unforeseen:** A Novel. By ALICE O'HANLON. New and Cheaper Edition. Post 8vo, illustrated boards, 2s.

**Oliphant (Mrs.) Novels by:**
- **Whiteladies.** With Illustrations by ARTHUR HOPKINS and H. WOODS. Crown 8vo, cloth extra, 3s. 6d.; post 8vo, illustrated boards, 2s.

Crown 8vo, cloth extra, 4s. 6d. each.
- **The Primrose Path.**
- **The Greatest Heiress in England.**

**O'Reilly.—Phœbe's Fortunes:** A Novel. With Illustrations by HENRY TUCK. Post 8vo, illustrated boards, 2s.

**O'Shaughnessy (Arth.), Works by:**
- **Songs of a Worker.** Fcap. 8vo, cloth extra, 7s. 6d.
- **Music and Moonlight.** Fcap. 8vo, cloth extra, 7s. 6d.
- **Lays of France.** Crown 8vo, cloth extra, 10s. 6d.

**Ouida, Novels by.** Crown 8vo, cloth extra, 5s. each; post 8vo, illustrated boards, 2s. each.

| | |
|---|---|
| Held in Bondage. | Signa. |
| Strathmore. | In a Winter City |
| Chandos. | Ariadne |
| Under Two Flags. | Friendship. |
| Cecil Castlemaine's Gage. | Moths. |
| Idalia. | Pipistrello. |
| Tricotrin. | A Village Commune. |
| Puck. | Bimbi. |
| Folle Farine. | In Maremma |
| TwoLittleWooden Shoes. | Wanda. |
| A Dog of Flanders. | Frescoes. |
| Pascarel. | Princess Napraxine. |

OUIDA, NOVELS BY, *continued*—

**Othmar**: A Novel. Cheaper Edition. Crown 8vo, cloth extra, 5s.

**Wisdom, Wit, and Pathos**, selected from the Works of OUIDA by F. SYDNEY MORRIS. Small crown 8vo, cloth extra, 5s.

**Page (H. A.), Works by:**

**Thoreau**: His Life and Aims: A Study. With a Portrait. Post 8vo, cloth limp, 2s. 6d.

**Lights on the Way**: Some Tales within a Tale. By the late J. H. ALEXANDER, B.A. Edited by H. A. PAGE. Crown 8vo, cloth extra, 6s.

**Animal Anecdotes.** Arranged on a New Principle. Crown 8vo, cloth extra, 5s. [*Shortly.*

**Parliamentary Elections and Electioneering in the Old Days (A History of).** Showing the State of Political Parties and Party Warfare at the Hustings and in the House of Commons from the Stuarts to Queen Victoria. Illustrated from the original Political Squibs, Lampoons, Pictorial Satires, and Popular Caricatures of the Time. By JOSEPH GREGO, Author of "Rowlandson and his Works," "The Life of Gillray," &c. Demy 8vo, cloth extra, with a Frontispiece coloured by hand, and nearly 100 Illustrations, 16s. One Hundred Large Paper Copies (each numbered) have also been prepared, price 32s. each.

**Pascal's Provincial Letters.** A New Translation, with Historical Introduction and Notes, by T. M'CRIE, D.D. Post 8vo, cloth limp, 2s.

**Patient's (The) Vade Mecum:** How to get most Benefit from Medical Advice. By WILLIAM KNIGHT, M.R.C.S., and EDWARD KNIGHT, L.R.C.P. Crown 8vo, 1s.; cloth, 1s. 6d.

**Paul Ferroll:**

Post 8vo, illustrated boards, 2s. each.

**Paul Ferroll**: A Novel.

**Why Paul Ferroll Killed his Wife.**

**Paul.—Gentle and Simple.** By MARGARET AGNES PAUL. With a Frontispiece by HELEN PATERSON. Cr. 8vo, cloth extra, 3s. 6d.; post 8vo, illustrated boards, 2s.

**Payn (James), Novels by.**

Crown 8vo, cloth extra, 3s. 6d. each post 8vo, illustrated boards, 2s. each.

**Lost Sir Massingberd.**
**The Best of Husbands.**
**Walter's Word.** | **Halves.**
**What He Cost Her.**
**Less Black than we're Painted.**
**By Proxy.** | **High Spirits.**
**Under One Roof.** | **Carlyon's Year.**

PAYN (JAMES), NOVELS BY, *continued*—

**A Confidential Agent.**
**Some Private Views.**
**A Grape from a Thorn.**
**For Cash Only.** | **From Exile.**
**Kit: A Memory.**
**The Canon's Ward.**

Post 8vo, illustrated boards, 2s. each.

**A Perfect Treasure.**
**Bentinck's Tutor.** | **Murphy's Master.**
**Fallen Fortunes.**
**A County Family.** | **At Her Mercy.**
**A Woman's Vengeance.**
**Cecil's Tryst.**
**The Clyffards of Clyffe.**
**The Family Scapegrace.**
**The Foster Brothers.**
**Found Dead.**
**Gwendoline's Harvest.**
**Humorous Stories.**
**Like Father, Like Son.**
**A Marine Residence.**
**Married Beneath Him.**
**Mirk Abbey.**
**Not Wooed, but Won.**
**Two Hundred Pounds Reward.**

**In Peril and Privation**: Stories of Marine Adventure Re-told. A Book for Boys. With numerous Illustrations. Crown 8vo, cloth gilt, 6s.

**The Talk of the Town**: A Novel. With Twelve Illustrations by HARRY FURNISS. Cr. 8vo, cl. extra, 3s. 6d.

**The Fly on the Wheel**: Humorous Papers. Crown 8vo, cloth extra, 6s. [*In the press.*

**Pears.—The Present Depression in Trade**: Its Causes and Remedies. Being the "Pears" Prize Essays (of One Hundred Guineas). By EDWIN GOADBY and WILLIAM WATT. With an Introductory Paper by Prof. LEONE LEVI, F.S.A., F.S.S. Demy 8vo, 1s.

**Pennell (H. Cholmondeley), Works by:**

Post 8vo, cloth limp, 2s. 6d. each.

**Puck on Pegasus.** With Illustrations.

**Pegasus Re-Saddled.** With Ten full-page Illusts. by G. DU MAURIER.

**The Muses of Mayfair.** Vers de Société, Selected and Edited by H. C. PENNELL.

**Phelps (E. Stuart), Works by:**

Post 8vo, 1s. each; cloth limp, 1s. 6d. each.

**Beyond the Gates.** By the Author of "The Gates Ajar."

**An Old Maid's Paradise.**

**Burglars in Paradise.** [*Shortly.*

**Pirkis (Mrs. C. L.), Novels by:**

**Trooping with Crows.** Fcap. 8vo, picture cover, 1s.

**Lady Lovelace.** Post 8vo, illustrated boards, 2s. [*Preparing.*

"SECRET OUT" SERIES, *continued*—

**The Pyrotechnist's Treasury; or,** Complete Art of Making Fireworks. By THOMAS KENTISH. With numerous Illustrations.

**The Art of Amusing:** A Collection of Graceful Arts, Games, Tricks, Puzzles, and Charades. By FRANK BELLEW. With 300 Illustrations.

**Hanky-Panky:** Very Easy Tricks, Very Difficult Tricks, White Magic Sleight of Hand. Edited by W. H. CREMER. With 200 Illustrations.

**The Merry Circle:** A Book of New Intellectual Games and Amusements. By CLARA BELLEW. Many Illusts.

**Magician's Own Book:** Performances with Cups and Balls, Eggs, Hats, Handkerchiefs, &c. All from actual Experience. Edited by W. H. CREMER. 200 Illustrations.

**Senior.—By Stream and Sea.** By WILLIAM SENIOR. Post 8vo, cloth l mp, 2s. 6d.

**Seven Sagas (The) of Prehistoric Man.** By JAMES H. STODDART, Author of "The Village Life." Crown 8vo, cloth extra, 6s.

**Shakespeare:**

**The First Folio Shakespeare.**—MR. WILLIAM SHAKESPEARE'S Comedies, Histories, and Tragedies. Published according to the true Originall Copies. London, Printed by ISAAC IAGGARD and ED. BLOUNT. 1623.—A Reproduction of the extremely rare original, in reduced facsimile, by a photographic process—ensuring the strictest accuracy in every detail. Small 8vo, half-Roxburghe, 7s. 6d.

**The Lansdowne Shakespeare.** Beautifully printed in red and black, in small but very clear type. With engraved facsimile of DROESHOUT'S Portrait. Post 8vo, cloth extra, 7s. 6d.

**Shakespeare for Children: Tales from Shakespeare.** By CHARLES and MARY LAMB. With numerous Illustrations, coloured and plain, by J. MOYR SMITH. Cr. 4to, cl. gilt, 6s.

**The Handbook of Shakespeare Music.** Being an Account of 350 Pieces of Music, set to Words taken from the Plays and Poems of Shakespeare, the compositions ranging rom the Elizabethan Age to the Present Time. By ALFRED ROFFE. 4to, half-Roxburghe, 7s.

**A Study of Shakespeare.** By ALGERNON CHARLES SWINBURNE. Crown 8vo, cloth extra. 8s.

**Shelley's Complete Works,** in Four Vols., post 8vo, cloth limp, 8s.; or separately, 2s. each. Vol. I. contains his Early Poems, Queen Mab, &c., with an Introduction by LEIGH HUNT; Vol. II., his Later Poems, Laon and Cythna, &c.; Vol. III., Posthumous Poems, the Shelley Papers, &c.: Vol. IV., his Prose Works, including A Refutation of Deism, Zastrozzi, St. Irvyne, &c.

**Sheridan:—**

**Sheridan's Complete Works,** with Life and Anecdotes. Including his Dramatic Writings, printed from the Original Editions, his Works in Prose and Poetry, Translations, Speeches, Jokes, Puns, &c. With a Collection of Sheridaniana. Crown 8vo, cloth extra, gilt, with 10 full-page Tinted Illustrations, 7s. 6d.

**Sheridan's Comedies: The Rivals, and The School for Scandal.** Edited, with an Introduction and Notes to each Play, and a Biographical Sketch of Sheridan, by BRANDER MATTHEWS. With Decorative Vignettes and 10 full-page Illusts. Demy 8vo, half-parchment, 12s. 6d.

**Short Sayings of Great Men.** With Historical and Explanatory Notes by SAMUEL A. BENT, M.A. Demy 8vo, cloth extra, 7s. 6d.

**Sidney's (Sir Philip) Complete Poetical Works,** including all those in "Arcadia." With Portrait, Memorial-Introduction, Notes, &c., by the Rev. A. B. GROSART, D.D. Three Vols., crown 8vo, cloth boards, 18s.

**Signboards: Their History.** With Anecdotes of Famous Taverns and Remarkable Characters. By JACOB LARWOOD and JOHN CAMDEN HOTTEN. Crown 8vo, cloth extra, with 100 Illustrations, 7s. 6d.

**Sims (George R.), Works by:**

**How the Poor Live.** With 60 Illusts. by FRED. BARNARD. Large 4to, 1s.

**Rogues and Vagabonds.** Post 8vo, illust. boards, 2s.; cloth limp, 2s. 6d.

**The Ring o' Bells.** Post 8vo, illust. bds., 2s.; cloth, 2s. 6d.

**Sketchley.—A Match in the Dark.** By ARTHUR SKETCHLEY. Post 8vo, illustrated boards, 2s.

**Slang Dictionary, The:** Etymological, Historical, and Anecdotal. Crown 8vo, cloth extra, gilt, 6s. 6d.

**Smith (J. Moyr), Works by:**

**The Prince of Argolis:** A Story of the Old Greek Fairy Time. Small 8vo, cloth extra, with 130 Illusts., 3s. 6d.

SMITH (J. MOYR), WORKS BY, *continued*—
**Tales of Old Thule.** With numerous Illustrations. Cr. 8vo, cloth gilt, 6s.
**The Wooing of the Water Witch:** A Northern Oddity. With numerous Illustrations. Small 8vo, cl. ex., 6s.

**Society in London.** By A FOREIGN RESIDENT. New and Cheaper Edition, Revised, with an Additional Chapter on SOCIETY AMONG THE MIDDLE AND PROFESSIONAL CLASSES. Crown 8vo, 1s.; cloth, 1s. 6d.

**Spalding.—Elizabethan Demonology:** An Essay in Illustration of the Belief in the Existence of Devils, and the Powers possessed by Them. By T. A. SPALDING, LL.B. Cr. 8vo, cl. ex., 5s.

**Spanish Legendary Tales.** By Mrs. S. G. C. MIDDLEMORE, Author of "Round a Posada Fire." Crown 8vo, cloth extra, 6s.

**Speight (T. W.), Novels by:**
**The Mysteries of Heron Dyke.** With a Frontispiece by M. ELLEN EDWARDS. Crown 8vo, cloth extra, 3s. 6d.; post 8vo, illustrated bds., 2s.
**A Barren Title.** Cr. 8vo, 1s.; cl., 1s. 6d.

**Spenser for Children.** By M. H. TOWRY. With Illustrations by WALTER J. MORGAN. Crown 4to, with Coloured Illustrations, cloth gilt, 6s.

**Staunton.—Laws and Practice of Chess;** Together with an Analysis of the Openings, and a Treatise on End Games. By HOWARD STAUNTON. Edited by ROBERT B. WORMALD. New Edition, small cr. 8vo, cloth extra, 5s.

**Stedman. — The Poets of America.** With full Notes in Margin, and careful Analytical Index. By EDMUND CLARENCE STEDMAN, Author of "Victorian Poets." Cr. 8vo, cl. ex., 9s.

**Sterndale.—The Afghan Knife:** A Novel. By ROBERT ARMITAGE STERNDALE. Cr. 8vo, cloth extra, 3s. 6d.; post 8vo, illustrated boards, 2s.

**Stevenson (R. Louis), Works by:**
**Travels with a Donkey in the Cevennes.** Fifth Ed. Frontispiece by W. CRANE. Post 8vo, cl. limp, 2s. 6d.
**An Inland Voyage.** With Front. by W. CRANE. Post 8vo, cl. lp., 2s. 6d.
**Virginibus Puerisque, and other Papers.** Crown 8vo, cloth extra, 6s.
**Familiar Studies of Men and Books.** Second Edit. Crown 8vo, cl. ex., 6s.
**New Arabian Nights.** Crown 8vo, cl. extra, 6s.; post 8vo, illust. bds., 2s.
**The Silverado Squatters.** With Frontispiece. Cr. 8vo, cloth extra, 6s. Cheap Edition, post 8vo, picture cover, 1s.; cloth, 1s. 6d.

STEVENSON (R. LOUIS), *continued*—
**Prince Otto:** A Romance. Fourth Edition. Crown 8vo, cloth extra, 6s.; post 8vo, illustrated boards, 2s.
**The Merry Men, and other Tales and Fables.** Cr. 8vo, cl. ex., 6s. [*Shortly*.

**St. John.—A Levantine Family.** By BAYLE ST. JOHN. Post 8vo, illustrated boards, 2s.

**Stoddard.—Summer Cruising in the South Seas.** By CHARLES WARREN STODDARD. Illust. by WALLIS MACKAY. Crown 8vo, cl. extra, 3s. 6d.

**Stories from Foreign Novelists.** With Notices of their Lives and Writings. By HELEN and ALICE ZIMMERN. Frontispiece. Crown 8vo, cloth extra, 3s. 6d.; post 8vo, illust. bds., 2s.

**St. Pierre.—Paul and Virginia, and The Indian Cottage.** By BERNARDIN ST. PIERRE. Edited, with Life, by Rev. E. CLARKE. Post 8vo, cl. lp., 2s.

**Strutt's Sports and Pastimes of the People of England;** including the Rural and Domestic Recreations, May Games, Mummeries, Shows, &c., from the Earliest Period to the Present Time. With 140 Illustrations. Edited by WILLIAM HONE. Crown 8vo, cloth extra, 7s. 6d.

**Suburban Homes (The) of London:** A Residential Guide to Favourite London Localities, their Society, Celebrities, and Associations. With Notes on their Rental, Rates, and House Accommodation. With Map of Suburban London. Cr. 8vo, cl. ex., 7s. 6d.

**Swift's Choice Works, in Prose and Verse.** With Memoir, Portrait, and Facsimiles of the Maps in the Original Edition of "Gulliver's Travels." Cr. 8vo, cloth extra, 7s. 6d.

**Swinburne (Algernon C.),** Works by:
**The Queen Mother and Rosamond.** Fcap. 8vo, 5s.
**Atalanta in Calydon.** Crown 8vo, 6s.
**Chastelard.** A Tragedy. Cr. 8vo, 7s.
**Poems and Ballads.** FIRST SERIES. Fcap. 8vo, 9s. Cr. 8vo, same price.
**Poems and Ballads.** SECOND SERIES. Fcap. 8vo, 9s. Cr. 8vo, same price.
**Notes on Poems and Reviews.** 8vo, 1s.
**Songs before Sunrise.** Cr. 8vo, 10s. 6d.
**Bothwell:** A Tragedy. Cr. 8vo, 12s. 6d.
**George Chapman:** An Essay. Crown 8vo, 7s.
**Songs of Two Nations.** Cr. 8vo, 6s.
**Essays and Studies.** Crown 8vo, 12s.
**Erechtheus:** A Tragedy. Cr. 8vo, 6s.
**Note of an English Republican on the Muscovite Crusade.** 8vo, 1s.

SWINBURNE'S (A. C.) WORKS, *continued*—
Note on Charlotte Bronte. Cr. 8vo, 6s.
A Study of Shakespeare. Cr. 8vo, 8s.
Songs of the Springtides. Cr. 8vo, 6s.
Studies in Song. Crown 8vo, 7s.
Mary Stuart: A Tragedy. Cr. 8vo, 8s.
Tristram of Lyonesse, and other Poems. Crown 8vo, 9s.
A Century of Roundels. Small 4to, 8s.
A Midsummer Holiday, and other Poems. Crown 8vo, 7s.
Marino Faliero: A Tragedy. Cr. 8vo, 6s.
A Study of Victor Hugo. Cr. 8vo, 6s.
Miscellanies. Crown 8vo, 12s.

**Symonds.—Wine, Women and Song:** Mediæval Latin Students' Songs. Now first translated into English Verse, with Essay by J. ADDINGTON SYMONDS. Small 8vo, parchment, 6s.

**Syntax's (Dr.) Three Tours:** In Search of the Picturesque, in Search of Consolation, and in Search of a Wife. With the whole of ROWLANDSON'S droll page Illustrations in Colours and a Life of the Author by J. C. HOTTEN. Med. 8vo, cloth extra, 7s. 6d.

**Taine's History of English Literature.** Translated by HENRY VAN LAUN. Four Vols., small 8vo, cloth boards, 30s.—POPULAR EDITION, Two Vols., crown 8vo, cloth extra, 15s.

**Taylor's (Bayard) Diversions of the Echo Club:** Burlesques of Modern Writers. Post 8vo, cl. limp, 2s.

**Taylor (Dr. J. E., F.L.S.), Works by.** Crown 8vo, cloth ex., 7s. 6d. each.
The Sagacity and Morality of Plants: A Sketch of the Life and Conduct of the Vegetable Kingdom. Coloured Frontispiece and 100 Illust.
Our Common British Fossils, and Where to Find Them: A Handbook for Students. With 331 Illustrations.

**Taylor's (Tom) Historical Dramas:** "Clancarty," "Jeanne Darc," "'Twixt Axe and Crown," "The Fool's Revenge," "Arkwright's Wife," "Anne Boleyn," "Plot and Passion." One Vol., cr. 8vo, cloth extra, 7s. 6d.
*** The Plays may also be had separately, at 1s. each.

**Tennyson (Lord): A Biographical Sketch.** By H. J. JENNINGS. With a Photograph-Portrait. Crown 8vo, cloth extra, 6s.

**Thackerayana: Notes and Anecdotes.** Illustrated by Hundreds of Sketches by WILLIAM MAKEPEACE THACKERAY, depicting Humorous Incidents in his School-life, and Favourite Characters in the books of his every-day reading. With Coloured Frontispiece. Cr. 8vo, cl. extra, 7s. 6d.

**Thomas (Bertha), Novels by:** Crown 8vo, cloth extra, 3s. 6d. each; post 8vo, illustrated boards, 2s. each.
Cressida. | Proud Maisie.
The Violin-Player.

**Thomas (M.).—A Fight for Life:** A Novel. By W. MOY THOMAS. Post 8vo, illustrated boards, 2s.

**Thomson's Seasons and Castle of Indolence.** With a Biographical and Critical Introduction by ALLAN CUNNINGHAM, and over 50 fine Illustrations on Steel and Wood. Crown 8vo, cloth extra, gilt edges, 7s. 6d.

**Thornbury (Walter), Works by**
Haunted London. Edited by EDWARD WALFORD, M.A. With Illustrations by F. W. FAIRHOLT, F.S.A. Crown 8vo, cloth extra, 7s. 6d.
The Life and Correspondence of J. M. W. Turner. Founded upon Letters and Papers furnished by his Friends and fellow Academicians. With numerous Illusts. in Colours, facsimiled from Turner's Original Drawings. Cr. 8vo, cl. extra, 7s. 6d.
Old Stories Re-told. Post 8vo, cloth limp, 2s. 6d.
Tales for the Marines. Post 8vo, illustrated boards, 2s.

**Timbs (John), Works by:**
Crown 8vo, cloth extra, 7s. 6d. each.
The History of Clubs and Club Life in London. With Anecdotes of its Famous Coffee-houses, Hostelries, and Taverns. With many Illusts.
English Eccentrics and Eccentricities: Stories of Wealth and Fashion, Delusions, Impostures, and Fanatic Missions, Strange Sights and Sporting Scenes, Eccentric Artists, Theatrical Folk, Men of Letters, &c. With nearly 50 Illusts.

**Trollope (Anthony), Novels by:**
Crown 8vo, cloth extra, 3s. 6d. each; post 8vo, illustrated boards, 2s. each.
The Way We Live Now.
Kept in the Dark.
Frau Frohmann. | Marion Fay.
Mr. Scarborough's Family.
The Land-Leaguers.

Post 8vo, illustrated boards, 2s. each.
The Golden Lion of Granpere.
John Caldigate. | American Senator

**Trollope (Frances E.), Novels by**
Crown 8vo, cloth extra, 3s. 6d. each; post 8vo, illustrated boards, 2s. each.
Like Ships upon the Sea.
Mabel's Progress. | Anne Furness.

**Trollope (T. A.).—Diamond Cut Diamond,** and other Stories. By T. ADOLPHUS TROLLOPE. Post 8vo, illustrated boards, 2s.

**Trowbridge.—Farnell's Folly:** A Novel. By J. T. TROWBRIDGE. Post 8vo, illustrated boards, 2s.

**Turgenieff. — Stories from Foreign Novelists.** By IVAN TURGENIEFF, and others. Cr. 8vo, cloth extra, 3s. 6d.; post 8vo, illustrated boards, 2s.

**Tytler (C. C. Fraser-). — Mistress Judith:** A Novel. By C. C. FRASER-TYTLER. Cr. 8vo, cloth extra, 3s. 6d.; post 8vo, illust. boards, 2s.

**Tytler (Sarah), Novels by:**
Crown 8vo, cloth extra, 3s. 6d. each; post 8vo, illustrated boards, 2s. each.
**What She Came Through.**
**The Bride's Pass.**
**Saint Mungo's City.**
**Beauty and the Beast.** With a Frontispiece by P. MACNAB.

Crown 8vo, cloth extra, 3s. 6d. each.
**Noblesse Oblige.** With Illustrations by F. A. FRASER.
**Citoyenne Jacqueline.** Illustrated by A. B. HOUGHTON.
**The Huguenot Family.** With Illusts.
**Lady Bell.** Front. by R. MACBETH.

**Buried Diamonds:** A Novel. Three Vols., crown 8vo.

**Van Laun.—History of French Literature.** By H. VAN LAUN. Three Vols., demy 8vo, cl. bds., 7s. 6d. each.

**Villari. — A Double Bond: A Story.** By LINDA VILLARI. Fcap. 8vo, picture cover, 1s.

**Walford (Edw., M.A.), Works by:**
**The County Families of the United Kingdom.** Containing Notices of the Descent, Birth, Marriage, Education, &c., of more than 12,000 distinguished Heads of Families, their Heirs Apparent or Presumptive, the Offices they hold or have held, their Town and Country Addresses, Clubs, &c. Twenty-sixth Annual Edition, for 1886, cloth gilt, 50s.
**The Shilling Peerage (1886).** Containing an Alphabetical List of the House of Lords, Dates of Creation, Lists of Scotch and Irish Peers, Addresses, &c. 32mo, cloth, 1s. Published annually.
**The Shilling Baronetage (1886).** Containing an Alphabetical List of the Baronets of the United Kingdom, short Biographical Notices, Dates of Creation, Addresses, &c. 32mo, cloth, 1s.

WALFORD'S (EDW.) WORKS, *continued*—
**The Shilling Knightage (1886).** Containing an Alphabetical List of the Knights of the United Kingdom, short Biographical Notices, Dates of Creation, Addresses, &c. 32mo, cl., 1s.
**The Shilling House of Commons (1886).** Containing a List of all the Members of Parliament, their Town and Country Addresses, &c. New Edition, embodying the results of the recent General Election. 32mo, cloth, 1s. Published annually.
**The Complete Peerage, Baronetage, Knightage, and House of Commons (1886).** In One Volume, royal 32mo, cloth extra, gilt edges, 5s.
**Haunted London.** By WALTER THORNBURY. Edited by EDWARD WALFORD, M.A. With Illustrations by F. W. FAIRHOLT, F.S.A. Crown 8vo, cloth extra, 7s. 6d.

**Walton and Cotton's Complete Angler;** or, The Contemplative Man's Recreation; being a Discourse of Rivers, Fishponds, Fish and Fishing, written by IZAAK WALTON; and Instructions how to Angle for a Trout or Grayling in a clear Stream, by CHARLES COTTON. With Original Memoirs and Notes by Sir HARRIS NICOLAS, and 61 Copperplate Illustrations. Large crown 8vo, cloth antique, 7s. 6d.

**Walt Whitman, Poems by.** Selected and edited, with an Introduction, by WILLIAM M. ROSSETTI. A New Edition, with a Steel Plate Portrait. Crown 8vo, printed on hand-made paper and bound in buckram, 6s.

**Wanderer's Library, The:**
Crown 8vo, cloth extra, 3s. 6d. each.
**Wanderings in Patagonia;** or, Life among the Ostrich-Hunters. By JULIUS BEERBOHM. Illustrated.
**Camp Notes:** Stories of Sport and Adventure in Asia, Africa, and America. By FREDERICK BOYLE.
**Savage Life.** By FREDERICK BOYLE.
**Merrie England in the Olden Time** By GEORGE DANIEL. With Illustrations by ROBT. CRUIKSHANK.
**Circus Life and Circus Celebrities.** By THOMAS FROST.
**The Lives of the Conjurers.** By THOMAS FROST.
**The Old Showmen and the Old London Fairs.** By THOMAS FROST.
**Low-Life Deeps.** An Account of the Strange Fish to be found there. By JAMES GREENWOOD.
**The Wilds of London.** By JAMES GREENWOOD.
**Tunis:** The Land and the People. By the Chevalier de HESSE-WARTEGG. With 22 Illustrations.

WANDERER'S LIBRARY, THE, *continued*—

**The Life and Adventures of a Cheap Jack.** By One of the Fraternity. Edited by CHARLES HINDLEY.

**The World Behind the Scenes.** By PERCY FITZGERALD.

**Tavern Anecdotes and Sayings:** Including the Origin of Signs, and Reminiscences connected with Taverns, Coffee Houses, Clubs, &c. By CHARLES HINDLEY. With Illusts.

**The Genial Showman:** Life and Adventures of Artemus Ward. By E. P. HINGSTON. With a Frontispiece.

**The Story of the London Parks.** By JACOB LARWOOD. With Illusts.

**London Characters.** By HENRY MAYHEW. Illustrated.

**Seven Generations of Executioners:** Memoirs of the Sanson Family (1688 to 1847). Edited by HENRY SANSON.

**Summer Cruising in the South Seas.** By C. WARREN STODDARD. Illustrated by WALLIS MACKAY.

## Warner.—A Roundabout Journey.

By CHARLES DUDLEY WARNER, Author of "My Summer in a Garden." Crown 8vo, cloth extra, 6s.

## Warrants, &c.:—

**Warrant to Execute Charles I.** An exact Facsimile, with the Fifty-nine Signatures, and corresponding Seals. Carefully printed on paper to imitate the Original, 22 in. by 14 in. Price 2s.

**Warrant to Execute Mary Queen of Scots.** An exact Facsimile, including the Signature of Queen Elizabeth, and a Facsimile of the Great Seal. Beautifully printed on paper to imitate the Original MS. Price 2s.

**Magna Charta.** An exact Facsimile of the Original Document in the British Museum, printed on fine plate paper, nearly 3 feet long by 2 feet wide, with the Arms and Seals emblazoned in Gold and Colours. Price 5s.

**The Roll of Battle Abbey;** or, A List of the Principal Warriors who came over from Normandy with William the Conqueror, and Settled in this Country, A.D. 1066-7. With the principal Arms emblazoned in Gold and Colours. Price 5s.

## Weather, How to Foretell the,

**with the Pocket Spectroscope.** By F. W. CORY, M.R.C.S. Eng., F.R.Met. Soc., &c. With 10 Illustrations. Crown 8vo, 1s.; cloth, 1s. 6d.

## Westropp.—Handbook of Pottery and Porcelain;

or, History of those Arts from the Earliest Period. By HODDER M. WESTROPP. With numerous Illustrations, and a List of Marks. Crown 8vo, cloth limp, 4s. 6d.

## Whistler's (Mr.) "Ten o'Clock."

Uniform with his "Whistler *v.* Ruskin: Art and Art Critics." Cr. 8vo, 1s. [*Shortly.*

## Williams (W. Mattieu, F.R.A.S.), Works by:

**Science Notes.** See the GENTLEMAN'S MAGAZINE. 1s. Monthly.

**Science in Short Chapters.** Crown 8vo, cloth extra, 7s. 6d.

**A Simple Treatise on Heat.** Crown 8vo, cloth limp, with Illusts., 2s. 6d.

**The Chemistry of Cookery.** Crown 8vo, cloth extra, 6s.

## Wilson (Dr. Andrew, F.R.S.E.), Works by:

**Chapters on Evolution:** A Popular History of the Darwinian and Allied Theories of Development. Third Edition. Crown 8vo, cloth extra, with 259 Illustrations, 7s. 6d.

**Leaves from a Naturalist's Notebook.** Post 8vo, cloth limp, 2s. 6d.

**Leisure-Time Studies,** chiefly Biological. Third Edit., with New Preface. Cr. 8vo, cl. ex., with Illusts., 6s.

**Studies in Life and Sense.** With numerous Illustrations. Crown 8vo, cloth extra, 6s. [*Preparing.*

**Common Accidents, and How to Treat them.** By Dr. ANDREW WILSON and others. With numerous Illustrations. Crown 8vo, 1s.; cloth limp, 1s. 6d.

## Winter (J. S.), Stories by:

**Cavalry Life.** Post 8vo, illust. bds., 2s.

**Regimental Legends.** Crown 8vo, cloth extra, 3s. 6d.; post 8vo, illustrated boards, 2s.

## Women of the Day: A Biographical Dictionary of Notable Contemporaries.

By FRANCES HAYS. Crown 8vo, cloth extra, 5s.

## Wood.—Sabina: A Novel.

By Lady WOOD. Post 8vo, illust. bds., 2s.

## Words, Facts, and Phrases:

A Dictionary of Curious, Quaint, and Out-of-the-Way Matters. By ELIEZER EDWARDS. New and cheaper issue, cr. 8vo, cl. ex., 7s. 6d.; half-bound, 9s.

## Wright (Thomas), Works by:

Crown 8vo, cloth extra, 7s. 6d. each.

**Caricature History of the Georges.** (The House of Hanover.) With 400 Pictures, Caricatures, Squibs, Broadsides, Window Pictures, &c.

**History of Caricature and of the Grotesque in Art, Literature, Sculpture, and Painting.** Profusely Illustrated by F. W. FAIRHOLT, F.S.A.

## Yates (Edmund), Novels by:

Post 8vo, illustrated boards, 2s. each.

**Castaway.** | **The Forlorn Hope.**

**Land at Last.**

## THREE-VOLUME NOVELS IN THE PRESS.

*WILKIE COLLINS'S NEW NOVEL.*

**The Evil Genius:** A Novel. By Wilkie Collins, Author of "The Woman in White." Three Vols., crown 8vo.

*WALTER BESANT'S NEW NOVEL.*

**Children of Gibeon:** A Novel. By Walter Besant, Author of "All Sorts and Conditions of Men," "Dorothy Forster," &c. Three Vols., crown 8vo.

*MRS. HUNT'S NEW NOVEL.*

**That other Person:** A Novel. By Mrs. Alfred Hunt, Author of "Thornicroft's Model," "The Leaden Casket," &c. Three Vols., crown 8vo.

*GRANT ALLEN'S NEW NOVEL.*

**In all Shades:** A Novel. By Grant Allen, Author of "Strange Stories," "Philistia," "Babylon," &c. Three Vols., crown 8vo.

*HALL CAINE'S NEW NOVEL.*

**A Son of Hagar:** A Novel. By T. Hall Caine, Author of "The Shadow of a Crime," &c. Three Vols., crown 8vo.

## THE PICCADILLY NOVELS.

Popular Stories by the Best Authors. Library Editions, many Illustrated, crown 8vo, cloth extra, 3s. 6d. each.

*BY MRS. ALEXANDER.*

Maid, Wife, or Widow?

*BY GRANT ALLEN.*

Philistia.

*BY BASIL.*

A Drawn Game.
"The Wearing of the Green."

*BY W. BESANT & JAMES RICE.*

Ready-Money Mortiboy.
My Little Girl.
The Case of Mr. Lucraft.
This Son of Vulcan.
With Harp and Crown
The Golden Butterfly.
By Celia's Arbour.
The Monks of Thelema.
'Twas in Trafalgar's Bay.
The Seamy Side.
The Ten Years' Tenant.
The Chaplain of the Fleet.

*BY WALTER BESANT.*

All Sorts and Conditions of Men.
The Captains' Room
All in a Garden Fair
Dorothy Forster.
Uncle Jack.

*BY ROBERT BUCHANAN.*

A Child of Nature.
God and the Man.
The Shadow of the Sword.
The Martyrdom of Madeline.
Love Me for Ever.
Annan Water. | The New Abelard.
Matt. | Foxglove Manor.
The Master of the Mine.

*BY HALL CAINE.*

The Shadow of a Crime.

*BY MRS. H. LOVETT CAMERON.*

Deceivers Ever. | Juliet's Guardian.

*BY MORTIMER COLLINS.*

Sweet Anne Page.
Transmigration.
From Midnight to Midnight.

*MORTIMER & FRANCES COLLINS.*

Blacksmith and Scholar.
The Village Comedy.
You Play me False.

*BY WILKIE COLLINS.*

Antonina.
Basil.
Hide and Seek.
The Dead Secret.
Queen of Hearts.
My Miscellanies.
Woman in White.
The Moonstone.
Man and Wife.
Poor Miss Finch.
Miss or Mrs.?
New Magdalen.
The Frozen Deep.
The Law and the Lady.
The Two Destinies
Haunted Hotel.
The Fallen Leaves
Jezebel's Daughter
The Black Robe.
Heart and Science
I Say No.

*BY DUTTON COOK.*

Paul Foster's Daughter.

*BY WILLIAM CYPLES*

Hearts of Gold

*BY ALPHONSE DAUDET.*

The Evangelist; or, Port Salvation.

*BY JAMES DE MILLE.*

A Castle in Spain.

*BY J LEITH DERWENT*

Our Lady of Tears. | Circe's Lovers

*BY M. BETHAM-EDWARDS.*

Felicia. | Kitty.

*BY MRS. ANNIE EDWARDES.*

Archie Lovell.

*BY R. E. FRANCILLON.*

Queen Cophetua. | A Real Queen.
One by One.

*Prefaced by Sir BARTLE FRERE.*

Pandurang Hari.

*BY EDWARD GARRETT.*

The Capel Girls.

PICCADILLY NOVELS, *continued*—

*BY CHARLES GIBBON.*

Robin Gray. | For Lack of Gold.
What will the World Say?
In Honour Bound.
Queen of the Meadow.
The Flower of the Forest.
A Heart's Problem.
The Braes of Yarrow.
The Golden Shaft. | Of High Degree.
Fancy Free. | Loving a Dream.
A Hard Knot. | Fancy Free.

*BY THOMAS HARDY.*

Under the Greenwood Tree.

*BY JULIAN HAWTHORNE.*

Garth. | Ellice Quentin.
Sebastian Strome.
Prince Saroni's Wife.
Dust. | Fortune's Fool.
Beatrix Randolph.
Miss Cadogna.
Love—or a Name.

*BY SIR A. HELPS.*

Ivan de Biron.

*BY MRS. CASHEL HOEY.*

The Lover's Creed.

*BY MRS. ALFRED HUNT.*

Thornicroft's Model.
The Leaden Casket.
Self-Condemned.

*BY JEAN INGELOW.*

Fated to be Free.

*BY HARRIETT JAY.*

The Queen of Connaught

*BY HENRY KINGSLEY.*

Number Seventeen.

*BY E. LYNN LINTON.*

Patricia Kemball.
Atonement of Leam Dundas.
The World Well Lost.
Under which Lord?
With a Silken Thread.
The Rebel of the Family
"My Love!" | Ione.

*BY HENRY W. LUCY.*

Gideon Fleyce.

*BY JUSTIN McCARTHY, M.P.*

The Waterdale Neighbours.
My Enemy's Daughter.
Linley Rochford. | A Fair Saxon.
Dear Lady Disdain.
Miss Misanthrope. | Donna Quixote
The Comet of a Season.
Maid of Athens.
Camiola.

*BY GEORGE MACDONALD.*

Paul Faber, Surgeon.
Thomas Wingfold, Curate.

*BY MRS. MACDONELL.*

Quaker Cousins.

PICCADILLY NOVELS, *continued*—

*BY FLORENCE MARRYAT.*

Open! Sesame! | Written in Fire

*BY D. CHRISTIE MURRAY.*

Life's Atonement. | Coals of Fire.
Joseph's Coat. | Val Strange.
A Model Father. | Hearts.
By the Gate of the Sea
The Way of the World.
A Bit of Human Nature.
First Person Singular.
Cynic Fortune.

*BY MRS. OLIPHANT.*

Whiteladies.

*BY MARGARET A. PAUL.*

Gentle and Simple.

*BY JAMES PAYN.*

Lost Sir Massingberd. | A Confidential Agent.
Best of Husbands | From Exile.
Halves. | A Grape from a Thorn.
Walter's Word. | For Cash Only.
What He Cost Her | Some Private Views.
Less Black than We're Painted. | Kit: A Memory.
By Proxy. | The Canon's Ward.
High Spirits. | The Talk of the Town.
Under One Roof.
Carlyon's Year.

*BY E. C. PRICE.*

Valentina. | The Foreigners
Mrs. Lancaster's Rival.

*BY CHARLES READE.*

It is Never Too Late to Mend.
Hard Cash.
Peg Woffington.
Christie Johnstone.
Griffith Gaunt. | Foul Play.
The Double Marriage.
Love Me Little, Love Me Long.
The Cloister and the Hearth.
The Course of True Love.
The Autobiography of a Thief.
Put Yourself in His Place.
A Terrible Temptation.
The Wandering Heir. | A Simpleton.
A Woman-Hater. | Readiana.
Singleheart and Doubleface.
The Jilt.
Good Stories of Men and other Animals.

*BY MRS. J. H. RIDDELL.*

Her Mother's Darling.
Prince of Wales's Garden-Party.
Weird Stories.

*BY F. W. ROBINSON.*

Women are Strange.
The Hands of Justice.

*BY JOHN SAUNDERS.*

Bound to the Wheel.
Guy Waterman.
Two Dreamers.
One Against the World.
The Lion in the Path.

CHEAP POPULAR NOVELS, *continued*—

WILKIE COLLINS, *continued.*

Man and Wife. | Haunted Hotel.
Poor Miss Finch. | The Fallen Leaves.
Miss or Mrs.? | Jezebel's Daughter
New Magdalen. | The Black Robe.
The Frozen Deep. | Heart and Science
Law and the Lady. | "I Say No."
The Two Destinies

*BY MORTIMER COLLINS.*

Sweet Anne Page. | From Midnight to Midnight.
Transmigration.
A Fight with Fortune.

*MORTIMER & FRANCES COLLINS.*

Sweet and Twenty. | Frances.
Blacksmith and Scholar.
The Village Comedy.
You Play me False.

*BY DUTTON COOK.*

Leo. | Paul Foster's Daughter.

*BY C. EGBERT CRADDOCK.*

The Prophet of the Great Smoky Mountains.

*BY WILLIAM CYPLES.*

Hearts of Gold.

*BY ALPHONSE DAUDET.*

The Evangelist; or, Port Salvation.

*BY JAMES DE MILLE.*

A Castle in Spain.

*BY J. LEITH DERWENT.*

Our Lady of Tears. | Circe's Lovers.

*BY CHARLES DICKENS.*

Sketches by Boz. | Oliver Twist.
Pickwick Papers. | Nicholas Nickleby

*BY MRS. ANNIE EDWARDES.*

A Point of Honour. | Archie Lovell

*BY M. BETHAM-EDWARDS.*

Felicia. | Kitty.

*BY EDWARD EGGLESTON.*

Roxy.

*BY PERCY FITZGERALD.*

Bella Donna. | Never Forgotten.
The Second Mrs. Tillotson.
Polly.
Seventy-five Brooke Street.
The Lady of Brantome.

*BY ALBANY DE FONBLANQUE.*

Filthy Lucre.

*BY R. E. FRANCILLON.*

Olympia. | Queen Cophetua.
One by One. | A Real Queen.

*Prefaced by Sir H. BARTLE FRERE.*

Pandurang Hari.

*BY HAIN FRISWELL.*

One of Two.

*BY EDWARD GARRETT.*

The Capel Girls.

CHEAP POPULAR NOVELS, *continued*—

*BY CHARLES GIBBON.*

Robin Gray. | The Flower of the Forest.
For Lack of Gold. | A Heart's Problem
What will the World Say? | The Braes of Yarrow.
In Honour Bound. | The Golden Shaft
In Love and War. | Of High Degree.
For the King. | Fancy Free.
In Pastures Green | By Mead and Stream.
Queen of the Meadow.

*BY WILLIAM GILBERT.*

Dr. Austin's Guests.
The Wizard of the Mountain.
James Duke.

*BY JAMES GREENWOOD.*

Dick Temple.

*BY ANDREW HALLIDAY.*

Every-Day Papers.

*BY LADY DUFFUS HARDY.*

Paul Wynter's Sacrifice.

*BY THOMAS HARDY.*

Under the Greenwood Tree.

*BY J. BERWICK HARWOOD.*

The Tenth Earl.

*BY JULIAN HAWTHORNE.*

Garth. | Sebastian Strome
Ellice Quentin. | Dust.
Prince Saroni's Wife.
Fortune's Fool. | Beatrix Randolph.

*BY SIR ARTHUR HELPS.*

Ivan de Biron.

*BY MRS. CASHEL HOEY.*

The Lover's Creed.

*BY TOM HOOD.*

A Golden Heart.

*BY MRS. GEORGE HOOPER.*

The House of Raby.

*BY VICTOR HUGO.*

The Hunchback of Notre Dame.

*BY MRS. ALFRED HUNT.*

Thornicroft's Model.
The Leaden Casket.
Self-Condemned.

*BY JEAN INGELOW.*

Fated to be Free.

*BY HARRIETT JAY.*

The Dark Colleen.
The Queen of Connaught.

*BY MARK KERSHAW.*

Colonial Facts and Fictions.

*BY HENRY KINGSLEY.*

Oakshott Castle.

*BY E. LYNN LINTON.*

Patricia Kemball.
The Atonement of Leam Dundas.
The World Well Lost.
Under which Lord?

CHEAP POPULAR NOVELS, *continued*—

LYNN LINTON, *continued*—

With a Silken Thread.
The Rebel of the Family.
"My Love
Ione.

*BY HENRY W. LUCY.*

Gideon Fleyce.

*BY JUSTIN McCARTHY, M.P.*

Dear Lady Disdain
The Waterdale Neighbours.
My Enemy's Daughter.
A Fair Saxon.
Linley Rochford.
Miss Misanthrope
Donna Quixote.
The Comet of a Season.
Maid of Athens.

*BY GEORGE MACDONALD.*

Paul Faber, Surgeon.
Thomas Wingfold, Curate.

*BY MRS. MACDONELL.*

Quaker Cousins.

*BY KATHARINE S. MACQUOID.*

The Evil Eye.
Lost Rose.

*BY W. H. MALLOCK.*

The New Republic.

*BY FLORENCE MARRYAT.*

Open! Sesame
A Harvest of Wild Oats.
A Little Stepson.
Fighting the Air
Written in Fire.

*BY J. MASTERMAN.*

Half-a-dozen Daughters.

*BY BRANDER MATTHEWS.*

A Secret of the Sea.

*BY JEAN MIDDLEMASS.*

Touch and Go.
Mr. Dorillion.

*BY D. CHRISTIE MURRAY.*

A Life's Atonement
A Model Father.
Joseph's Coat.
Coals of Fire.
By the Gate of the Sea.
Val Strange.
Hearts.
The Way of the World.
A Bit of Human Nature.

*BY ALICE O'HANLON.*

The Unforeseen.

*BY MRS. OLIPHANT.*

Whiteladies.

*BY MRS. ROBERT O'REILLY.*

Phœbe's Fortunes.

*BY OUIDA.*

Held in Bondage.
Strathmore.
Chandos.
Under Two Flags.
Idalia.
Cecil Castlemaine's Gage.
Tricotrin
Puck.
Folle Farine.
A Dog of Flanders.
Pascarel.
Signa.
Two Little Wooden Shoes.
In a Winter City.
Ariadne.
Friendship.
Moths.
Pipistrello.
A Village Commune.
Bimbi.
In Maremma.
Wanda.
Frescoes.
Princess Napraxine.

CHEAP POPULAR NOVELS, *continued*—

*BY MARGARET AGNES PAUL.*

Gentle and Simple.

*BY JAMES PAYN.*

Lost Sir Massingberd.
A Perfect Treasure.
Bentinck's Tutor.
Murphy's Master.
A County Family.
At Her Mercy.
A Woman's Vengeance.
Cecil's Tryst.
Clyffards of Clyffe
The Family Scapegrace.
Foster Brothers.
Found Dead.
Best of Husbands.
Walter's Word.
Halves.
Fallen Fortunes.
What He Cost Her
Humorous Stories
Gwendoline's Harvest.
£200 Reward.
Like Father, Like Son.
A Marine Residence.
Married Beneath Him.
Mirk Abbey.
Not Wooed, but Won.
Less Black than We're Painted.
By Proxy.
Under One Roof.
High Spirits.
Carlyon's Year.
A Confidential Agent.
Some Private Views.
From Exile.
A Grape from a Thorn.
For Cash Only.
Kit: A Memory.
The Canon's Ward

*BY EDGAR A. POE.*

The Mystery of Marie Roget.

*BY E. C. PRICE.*

Valentina.
The Foreigners.
Mrs. Lancaster's Rival.
Gerald.

*BY CHARLES READE.*

It is Never Too Late to Mend
Hard Cash.
Peg Woffington.
Christie Johnstone.
Griffith Gaunt.
Put Yourself in His Place.
The Double Marriage.
Love Me Little, Love Me Long.
Foul Play.
The Cloister and the Hearth.
The Course of True Love.
Autobiography of a Thief.
A Terrible Temptation.
The Wandering Heir.
A Simpleton.
A Woman-Hater.
Readiana.
The Jilt.
Singleheart and Doubleface.
Good Stories of Men and other Animals.

*BY MRS. J. H. RIDDELL.*

Her Mother's Darling.
Prince of Wales's Garden Party.
Weird Stories.
The Uninhabited House.
Fairy Water.
The Mystery in Palace Gardens.

*BY F. W. ROBINSON.*

Women are Strange.
The Hands of Justice.

CHEAP POPULAR NOVELS, *continued*—

*BY JAMES RUNCIMAN.*

Skippers and Shellbacks.
Grace Balmaign's Sweetheart.

*BY W. CLARK RUSSELL.*

Round the Galley Fire.
On the Fo'k'sle Head.

*BY BAYLE ST. JOHN.*

A Levantine Family.

*BY GEORGE AUGUSTUS SALA.*

Gaslight and Daylight.

*BY JOHN SAUNDERS.*

Bound to the Wheel.
One Against the World.
Guy Waterman.
The Lion in the Path.
Two Dreamers.

*BY KATHARINE SAUNDERS.*

Joan Merryweather.
Margaret and Elizabeth.
The High Mills.

*BY GEORGE R. SIMS.*

Rogues and Vagabonds.
The Ring o' Bells.

*BY ARTHUR SKETCHLEY.*

A Match in the Dark.

*BY T. W. SPEIGHT.*

The Mysteries of Heron Dyke.

*BY R. A. STERNDALE.*

The Afghan Knife.

*BY R. LOUIS STEVENSON.*

New Arabian Nights.
Prince Otto.

*BY BERTHA THOMAS.*

Cressida. | Proud Maisie.
The Violin-Player.

*BY W. MOY THOMAS.*

A Fight for Life.

*BY WALTER THORNBURY.*

Tales for the Marines.

*BY T. ADOLPHUS TROLLOPE.*

Diamond Cut Diamond.

*BY ANTHONY TROLLOPE.*

The Way We Live Now.
The American Senator.
Frau Frohmann.
Marion Fay.
Kept in the Dark.
Mr. Scarborough's Family.
The Land-Leaguers.
The Golden Lion of Granpere.
John Caldigate.

*By FRANCES ELEANOR TROLLOPE*

Like Ships upon the Sea.
Anne Furness.
Mabel's Progress.

*BY J. T. TROWBRIDGE.*

Farnell's Folly.

CHEAP POPULAR NOVELS, *continued*—

*BY IVAN TURGENIEFF, &c.*

Stories from Foreign Novelists.

*BY MARK TWAIN.*

Tom Sawyer.
A Pleasure Trip on the Continent of Europe.
A Tramp Abroad.
The Stolen White Elephant.
Huckleberry Finn.

*BY C. C. FRASER-TYTLER.*

Mistress Judith.

*BY SARAH TYTLER.*

What She Came Through.
The Bride's Pass.
Saint Mungo's City.
Beauty and the Beast.

*BY J. S. WINTER.*

Cavalry Life. | Regimental Legends.

*BY LADY WOOD.*

Sabina.

*BY EDMUND YATES.*

Castaway. | The Forlorn Hope.
Land at Last.

*ANONYMOUS.*

Paul Ferroll.
Why Paul Ferroll Killed his Wife.

## POPULAR SHILLING BOOKS.

Jeff Briggs's Love Story. By BRET HARTE.
The Twins of Table Mountain. By BRET HARTE.
Mrs. Gainsborough's Diamonds. By JULIAN HAWTHORNE.
Kathleen Mavourneen. By Author of "That Lass o' Lowrie's."
Lindsay's Luck. By the Author of "That Lass o' Lowrie's."
Pretty Polly Pemberton. By the Author of "That Lass o' Lowrie's."
Trooping with Crows. By Mrs. PIRKIS.
The Professor's Wife. By LEONARD GRAHAM.
A Double Bond. By LINDA VILLARI.
Esther's Glove. By R. E. FRANCILLON.
The Garden that Paid the Rent. By TOM JERROLD.
Curly. By JOHN COLEMAN. Illustrated by J. C. DOLLMAN.
Beyond the Gates. By E. S. PHELPS.
An Old Maid's Paradise. By E. S. PHELPS.
Burglars in Paradise. By E. S. PHELPS.
Doom: An Atlantic Episode. By JUSTIN H. MACCARTHY, M.P.
Our Sensation Novel. Edited by JUSTIN H. MACCARTHY, M.P.
A Barren Title. By T. W. SPEIGHT.
The Silverado Squatters. By R. LOUIS STEVENSON.

J. OGDEN AND CO., PRINTERS, 29, 30 AND 31, GREAT SAFFRON HILL, E.C.